TECHNOLOGY MANAGEMENT IN ORGANIZATIONS

The author gratefully acknowledges the financial support of the
Burns Foods Endowment Fund, Faculty of Management,
the University of Lethbridge.

TECHNOLOGY MANAGEMENT IN ORGANIZATIONS

Urs E. Gattiker

HD
45
.G36
1990
West

SAGE PUBLICATIONS
The International Professional Publishers
Newbury Park London New Delhi

Copyright © 1990 by Sage Publications, Inc.

All rights reserved. No part of this book may be reproduced or utilized in any form or by any means, electronic or mechanical, including photocopying, recording, or by any information storage and retrieval system, without permission in writing from the publisher.

For information address:

SAGE Publications, Inc.
2111 West Hillcrest Drive
Newbury Park, California 91320

SAGE Publications Ltd.
28 Banner Street
London EC1Y 8QE
England

SAGE Publications India Pvt. Ltd.
M-32 Market
Greater Kailash I
New Delhi 110 048 India

Printed in the United States of America

Library of Congress Cataloging-in-Publication Data

Gattiker, Urs E.
    Technology management in organizations / by Urs E. Gattiker.
       p.   cm.
    Includes bibliographical references.
    ISBN 0-8039-3607-9. -- ISBN 0-8039-3608-7 (pbk.)
    1. Technological innovations--Management. 2. Manpower planning,
3. Organizational behavior. 4. Organizational change. I. Title.
HD45.G36   1990   90-8227
658.5'14--dc20   CIP

FIRST PRINTING, 1990

Sage Production Editor: Susan McElroy

Contents

Foreword	8
List of Abbreviations	13
Part I: Innovation and Technology-Induced Organizational Change	15
1. The Nature of Innovation	17
2. An Integrated Model of the Contingency-Culture Approach	28
3. Strategic Choice and Environmental Determinism	40
4. Determinism and Organizational Characteristics	57
Part II: Organizational Systems Affecting Technology Acquisition	73
5. Internal Labour Market Determinism and Strategic Human Resource Management	75
6. Employment Strategies, Recruitment, and Selection	93
7. Socialization and Appraisal Systems	114
8. Career Development and Training Systems	133
9. Compensation Systems	152
Part III: Cultural and Individual Factors Influencing Technology Adoption	173
10. The Cultural Context and the Individual	175
11. Inequality and Technology Acceptance	205
12. Employee Skills and Technology Management	228
13. Development of a Theory	256

14. Conclusion	280
Glossary	297
References	307
Author Index	325
Subject Index	332
About the Author	339

List of Figures

1.1	Classifying Technological Innovation	21
2.1	Technology Acquisition and Technology-Induced Organizational Adaptation as Seen Within Its Cultural and Environmental Context Using Both the Objective-Subjective and the Macro-Micro Dimensions	32
3.1	Relationship of Strategic Choice, Environmental Determinism in Technology Acquisition, and Organizational Adaptation by the Firm	46
5.1	Relationship of Choice in Strategic Human Resource Management and Internal Labour Market Determinism	81
11.1	Magnitude of Acceptance of or Resistance to Technology-Induced Organizational Change	215
12.1	A Matrix Approach to Job Skills and Job Knowledge	233
13.1	Office Work and Technology-Induced Organizational Adaptation: A Conceptual Model	263

List of Tables

3.1	Four Organizational Types: Approaches to Technology Acquisition and Organizational Adaptation	49
4.1	Four Organizational Types: General Characteristics	64
4.2	Four Organizational Types: Internal Characteristics	67
5.1	Four Human Resource Strategy and Internal Labour Market Types: Organizational Systems	84

6.1	Four Human Resource Strategy and Internal Labour Market Types: Employment Strategies	101
6.2	Four Human Resource Strategy and Internal Labour Market Types: Recruitment and Selection Systems	108
7.1	Four Human Resource Strategy and Internal Labour Market Types: Socialization, Appraisal, and Career Development Systems	127
8.1	Four Human Resource Strategy and Internal Labour Market Types: Career Development and Training Systems	147
9.1	Four Human Resource Management Strategy and Internal Labour Market Types: Compensation Packages	166
11.1	Organizational Culture and Technological Adaptation	225
12.1	Four Human Resource Management Strategy and Internal Labour Market Types: End User Knowledge and Skills	248

Foreword

As the economy of the 1980s and 1990s forces employers to be more conscious about technology than ever before, the success or failure of businesses and other organizations may hinge on decisions regarding technology management. And as the costs of doing business become increasingly linked to the technological needs of organizations (e.g., hardware, software, and employee training costs), people who make technology management decisions will bear a major responsibility for cost control and effective use of technology. Society too has turned its attention to the issue of new technology and, through unions and the legal system, will hold employers accountable for job safety, security, and quality of work life for employees as technology-induced changes occur. All this makes the technology management field dynamic and exciting and personally interesting as it affects each of us. The enthusiast will frequently ponder a firm's potential for successful technology management, or wonder why certain groups of employees resist further penetration of technology in their workplace, or why certain firms are always riding the crest of the technological wave while others lag behind.

This book attempts to provide a coherent introduction to the study of technology management in organizations, with a primarily human resource focus. This is no mean or simple task. The area has developed rapidly during the past decade; it is the scene of much effort and activity and, probably as a consequence, of much controversy and confusion. Conceptual schemes, typologies, and theories are generated with accelerating frequency but rather than being evaluated or sorted out they are simply added to a growing pile. Empirical studies abound—case studies of single or multiple situations, comparative studies of similar or diverse units, and, increasingly, longitudinal or panel studies of organizations managing technology-induced change. The result, however, merely augments our information without adding to our knowledge. When a field of study grows rapidly, we expect to see increasing signs of differentiation and fragmentation. Our expectations are confirmed in the case of technology management. The past decade has produced a disparate set of specialized approaches. Technology management has been used as part of corporate strategy, of an information or technology-focused system, or of research and development efforts and innovation management. Textbooks mirror the incoherent diversity of this explosive domain. The number of approaches, studies, books, journals, and fads increases, but we do not seem to realize a commensurate gain in our understanding.

Technology Management in Organizations was written, in one sense, to acknowledge this diversity, and to present the major dimensions underlying this complex field. Three theoretical perspectives are used to introduce some order into the existing chaos: organizational behaviour, organization theory, and end-user technology management. I have employed more than one perspective because I believe that no single existing perspective permits adequate comprehension of the important features and dimensions of technology management. I have used only three perspectives because I believe that most of the models and approaches currently used may be viewed as variants of one of these more general types. The basic "point of view" taken in this book is that technology management is human resource driven; the employees making up the organization have a vital role to play.

Why is a human resource focus grounded in organizational behaviour used in this book? Up to now, a great deal of research has been done that has produced, in many instances, little or no advancement of knowledge. It is reasonable to assume that the problem is not with the talent or motivation of those seeking to understand technology management but with the point of view, the perspective, the approach to analysis that has tended to dominate the field. This argument requires three elements for demonstration.

First, it is necessary to survey or overview the field of innovation management and technological change to point out what has been, or remains to be, accomplished. However, this book is not primarily an overview or a fully comprehensive review of the literature. It will instead focus attention upon those things underlying the processes of inquiry that have been pursued: innovation, technology management, information systems, and logistics.

Second, as the literature is discussed, we should develop a framework to help explain technology management within the social context in which it has developed. Thus the second objective is to provide some sense of the social context and its effects in this field.

Third, this book is devoted to the development of a human resource approach to the analysis of technology management. My intent is not to endorse a narrowly parochial perspective; one of the most attractive facets of the contemporary scene is the increasingly multidisciplinary character of research on technology management. The goal, rather, is to nudge technology management research in some new directions. My objective is to incorporate input from neighbouring disciplines into my essentially organizational behaviour approach. I believe that organizational behaviour theory is in the best position to integrate and transcend the more specialized contributions arising from alternative points of view.

I further believe that effective technology management decisions require a knowledge of both current theory and research regarding technology management and specific practice. Theory, research, and practice have all changed in the last decade as rapid technological change necessitates an ever-increasing understanding of the technology management process. In *Technology Management in Organizations*, theory and research will be presented and their link to current practices will be outlined whenever possible.

Although it might appear at first glance that this book draws primarily on North American references, this is not actually the case; the literature cited is mostly in English, however, to allow readers to locate and utilize it more easily. The exception to this general rule are the German references, included because of the prominence of the Federal Republic of Germany (FRG) — often referred to as West Germany in the technology field — and because they are accessible to me due to my knowledge of German. Still, there appear to be more North Americans than Europeans researching in this area, and the reference list is necessarily reflective of this phenomenon. To balance this, I have also used international examples, including ones from France, Great Britain, Italy, Japan, Korea, the Netherlands, New Zealand, Sweden, Switzerland and West Germany. Third World countries supply no examples because their situations are rarely comparable or applicable to those of the developed countries that concern themselves with innovative technology (Warner, 1987). Essentially, this book strives to attain an international focus, one that is grounded in North American thinking but that tries to integrate other models and concepts sometimes disregarded by North American researchers.

Part I of the book describes what is meant by technological innovation, acquisition of technology, and technology-induced organizational adaptation. We'll use a contingency framework, with the focus on the environmental factors influencing the degree of choice experienced by the organization when deciding whether to adopt new technology and make the changes necessary to use it effectively. I have developed a typology that locates a firm in one of four quadrants based on the level of choice in its strategic human resource management and the level of internal labour market determinism. Included is a description of the various general and internal characteristics inherent in a firm located in a particular quadrant.

Part II of this book focuses especially upon the internal processes that may affect technology-induced organizational change. Chapters 5 to 8 outline employment strategies, recruitment and selection systems, socialization and appraisal systems, career development and training systems, and compensation systems using the typology developed in Chapters 3 and 4. Also included is a description of the interrelationship between objective factors (environment, choice) and organizational systems in the human resource domain.

Part III builds upon this approach by studying both macro and micro dimensions of the technology acquisition and technology-induced organizational change process, concentrating upon the more subjective side of technology management — the perceptions, beliefs, and attitudes of the individual employee. My assumption is that, based upon the firm's strategic choice, environmental determinism, strategic human resource management, and internal labour market determinism, the organization develops its own system of values and beliefs that are passed on to its employees. The culture thus created is further reinforced by human resource systems (see Part II) that attract certain individuals to work for the firm.

Both the human resource systems and the current employees' values and attitudes towards technology must be considered in order to reduce potential resistance by the work force. Part III further outlines how cross-national differences may affect technology management for international firms and how perceptions of inequality may increase resistance. The approach presented here uses four categories on the basis of two dimensions: (a) level of criterion (i.e., macro and micro level of analysis) and (b) objectivity-subjectivity (i.e., quantifiable efficiency measures and effectiveness indicators such as employee perceptions).

Anyone reading this book in one sitting might point to some repetitiveness from chapter to chapter. I believe, however, that repeating a point at a later stage to improve clarity is "reader-friendly" rather than tedious. Even more important, it also means that individual chapters may be read as needed, as separate entities. I trust that you will agree.

Acknowledgments

Many persons have contributed to my own development as a social scientist and student of organizations, and the publication of a book such as this provides a welcome occasion to express my indebtedness and appreciation. At the Claremont Graduate School, Laurie Larwood introduced me to organizational behaviour in such a manner as to enlist my loyalties for the duration, and she taught me that good research could (and should) be both academically rigorous and socially useful. Barbara A. Gutek provided me with my first systematic exposure to survey research and gave me my first taste of field research in an organizational setting. Chester A. Schriesheim made me aware how important it is to research areas in which one has a personal interest and pursue such interests with vigour, imagination, and persistence. Jeffrey Pfeffer was instrumental in opening my eyes to critical analysis of organizational behaviour and theory and inciting me to challenge my own research and its contribution to knowledge.

At the University of Lethbridge, where I have worked for the last five years, I have had the good fortune to be associated with many great colleagues. Hiroshi Tanaka was probably most influential in getting me to write down my ideas, reflections and questions in book format. George Lermer, Dean of the Faculty of Management, provided the support and encouragement that helped the writing process along. I cannot name all those colleagues and students from whom I have learned but I must not fail to acknowledge the contribution to my education made by Shamsul Alam, Aaron Cohen, Cynthia Cunningham, Geoff England, Angela Hlavka, Valorie Hoye, Cathy Kirchmeyer, Veronica Lees, Bob Loo, Ken Nicol, Dan Paulson, Anil Pereira, and Bonny West.

I owe special thanks to the reviewers, who generously and thoughtfully provided excellent comments on earlier versions of this manuscript. They sometimes saw what I was doing better than I did, and they noted problems in the argument, in the literature covered, or in the presentation, which I have tried to correct. If I could be certain this manuscript was of the same quality as their comments, I would be pleased indeed.

My greatest appreciation goes to my three editorial assistants Brenda McPhail, HelenJane Shawyer, and Larry Steinbrenner, who were involved at different stages during the process of writing this book. Their insistence on clarity and logic in my writing has greatly improved the final product, although ultimate responsibility lies with me. Barb Driscoll, Karen Eriksen, Stella Kedoin, and Marlene Lapointe provided competent and timely word-processing assistance. Last but not least, I would like to acknowledge my wife Rosemarie's unfailing support and encouragement, and my daughter Melanie's patience with her daddy, who often did not have enough time to play.

List of Abbreviations

AAWU	American Automobile Workers Union
ACP	Airline control program
ATM	Automated teller machine
BARS	Behaviourally anchored rating scales
BIGA	Bundesamt für Industrie, Gewerbe und Arbeit (Swiss Federal Office for Industry, Trades and Work)
CAD	Computer-aided design
CCDO	*Canadian Classification and Dictionary of Occupations*
CNC	Computer-numeric control
CPA	Chartered public accountant
DP	Deutsche Bundespost (West German mail, telephone, telegraph, and videotex)
EDPS	Employee development planning system
EEO	Equal employment opportunity
EUTM	End-user technology management
Fax	Facsimile
Galileo	Airline reservation system (owned by British Airways, Swissair, United Airlines, and others)
GM	General Motors
HRM	Human resource management
IBM-PC	International Business Machines – personal computer
IG-Metall	German Metal Workers' Union
ILM	Internal labour market
ILMD	Internal labour market determinism
LAN	Local-Area Network
MBA	Master of business administration
NZZ	Neue Zürcher Zeitung

OECD	Organization for Economic Cooperation and Development
SABRE	Semi-Automated Business Reservation Environment (Airline reservation system owned by American Airlines)
SEAT	Volkswagen's Spanish subsidiary
SHRM	Strategic human resource management
SIF	Swedish Metal Workers' Union
SWISS PTT	Swiss mail: mail, telephone, telegraph, and videotex
TCO	The Central Organization of Salaried Employees (TCO) in Sweden
UBS	Union Bank of Switzerland
UPS	United Parcel Service
VDA	Verband der Automobilindustrie (German Association of Car Manufacturers)
VDT	Visual display terminal

PART I

Innovation and Technology-Induced Organizational Change

Innovation, including its successful implementation into production processes, has been heralded as the industrialized country's only avenue to sustaining economic growth and material wealth. In the past 20 years, countless articles and books have been written on the topic of successfully managing new technology and innovation in an organizational setting. Unfortunately, most research has concentrated on the innovation process itself and has neglected to study how firms cope with the changes inherent in introducing new technology into an organization. The innovation process is one side of the coin, but innovation-induced organizational change is the other; for the "innovation coin" to have value, or for successful acquisition and effective use of the innovation, the coin must be two-sided.

This book primarily concentrates on the processes of acquisition and technology-induced organizational change required to implement technological innovations successfully. We discuss organizations and their attempts to adopt technological innovations — especially computer and information system-related ones — and examine the changes needed to make effective use of new technology and assure consistency between the technology and internal processes and structures. Adopting innovations forces organizations to confront a variety of problems they would not normally have to deal with in their ongoing struggle with environmental factors. In fact, it is often more accurate to say that an organization is adapting the innovation or new technology to its internal structures and processes than the reverse. Identifying the problems created by incorporating innovation, and studying the different ways in which organizations may choose to handle them, is a priority if we are to understand the methods by which new technology is adopted and how a firm may cope with innovation-induced organizational change. The organization's approach to these problems will be affected both by choices made within the organization and by the influence of outside factors. Therefore, it is also necessary for us to try to determine the level of organizational choice and environmental determinism that we can expect to result in the kinds of structures, remuneration policies, and promotional ladders that make it easier for some organizations to adapt than others.

Part I of this book specifically focuses upon environmental factors' influence upon technology aquisition and technology-induced organizational change. For instance, environmental constraints such as government regulations (e.g., in such areas as toxic waste handling, market regulation, tariffs, and occupational health and safety) affect the choices and opportunities a firm may have to change its strategy. Chapters 1 through 4 will thus try to accomplish three things: (a) discussing and defining numerous terms used in the literature; (b) addressing the issue of organizational change and innovation in general, and technological innovation in particular; and (c) presenting a typology that should permit classification of organizations into four quadrants, dependent upon their level of organizational choice and environmental determinism.

The typology will be used to outline how placing an organization in a particular quadrant may help us understand differing approaches to technology acquisition and technology-induced organizational change. Moreover, we'll see the role played by the type of innovation introduced into the firm and the degree of market diffusion of newly adopted technologies. Part I also discusses how organizational choice and environmental determinism levels may affect general organizational characteristics such as ownership, governance, distinctive competence, and structure. Furthermore, we'll discuss the internal characteristics of firms in the typology's four quadrants, such as nature of task, organization of work, and communication systems.

Chapter 1 endeavours to explain some of the different forms of innovation that a company may face within the context of its environment. This chapter will provide us with working definitions of *innovation* and *technological innovation*, and clarify what types of innovation organizations should incorporate to remain successful. Chapter 2 presents a model combining the contingency and cross-cultural approaches to technology acquisition and technology-induced organizational change, and Chapter 3 broadly examines the process by which organizations absorb technological innovations. The chapter classes organizations according to four types, each experiencing varying degrees of environmental determinism and freedom of choice regarding their operations. And it outlines the effect of organization type upon the kind of innovation adopted by the firm and the way in which it is introduced. Chapter 4 discusses how organizational characteristics and structures may differ based on the four organizational types developed in the previous chapter.

This typology can be used to determine some principles that are helpful in explaining how organizations experience change when adopting technological innovation. We shall also address the problem of why, with so much effort expended upon studying technology acquisition, technology-induced organizational change, and the systems of classification that have been proposed to accompany them, previous typologies have not been as helpful in the study of technology management as in other fields of inquiry.

1. The Nature of Innovation

The central thesis of this book is that, to understand technology management, or an organization's success or failure in this domain, we must understand its context—that is, the ecology of the organization and the processes needed to facilitate innovation, adoption of new technology, and technology-induced organizational change. Innovation imposed upon organizations without internal receptivity is bound to fail. A firm that lacks the appropriate structure, processes, human resource systems, or change mechanisms will be unable to reconcile the innovation, the organization, the employees, and the desired outcome.

Innovations and their acquisition may not succeed because of internal resistance (e.g., Kimberly, 1981) resulting from incompatibilities between the nature of innovation and the existing configuration of interests and resources. The literature tends to assume that innovation acquisition is conducted in a rational and purposeful manner, with conscious forethought about the goals this acquisition is designed to achieve (Pfeffer, 1982, p. 7). But is this the case?

Analysis of innovation and its acquisition, and the changes this acquisition causes in an organization, requires a broad approach. This chapter has been written not only to review existing knowledge but also to identify important research areas in innovation studies, especially managerial and scholarly inquiries concerning technology. Its primary objective is to begin mapping the terrain by outlining various approaches to the study of innovation, its acquisition, and innovation-induced organizational change.

The Process of Innovation

What is *technological innovation*? As we shall see, it is often a part of many things with which it is not normally associated. *Innovation* itself is a rather broad concept that can be understood in a variety of ways. An innovation may be a new way of thinking or an invention that is the product of this way of thinking. In an organizational context, the term may have several meanings and has been variously understood to mean a disposition towards new ways of thinking or new ways of doing things, a willingness to develop new methods through research and development (R&D) within the organization itself, or a willingness to acquire new products, services, and processes from either inside or outside the organization. Innovative activity has differed and has had varying results through the course of human history. Rosegger (1986), describing innovation from a historical perspective, examined the nature of changes and identified major themes of innovative activity as dominating certain phases of long-term development. *Mechanization* from the late eighteenth and early nineteenth centuries involved a growing reliance upon machine power as opposed to human effort, and *standardization* and *specialization* made possible the economies of scale necessary for mass production. *Automation of production* brought the feedback-control principle to mechanized manufacturing and reduced the degree of human decision making necessary to keep the production process going. *Computerization* raised this principle to the level where the integration *and* coordination of sequential production stages became feasible. *Computer-aided manufacturing* is the latest step in this evolution, allowing production plants to run according to preprogrammed instructions with a minimum number of workers.

Some authors seem to equate R&D with innovation (Link & Tassey, 1987, pp. 5-10), but the R&D process will not necessarily lead to innovations that are ultimately marketable. Therefore, it seems an unlikely key to a definition of innovation. Research indicates that, compared with companies that have low R&D expenditures, firms that have high R&D expenditures are no more likely to be profitable (Hitt, Ireland, & Goryunov, 1988), yet profitability is something we might expect from a successfully innovative organization.

Although most study of innovation has concentrated on manufacturing, the greatest potential impact of technology may actually lie in how it alters the way business is done in the service sector of the economy. This sector is made up of organizations that exist for the purpose of providing specialized services such as transportation, real estate, banking, finance, and public assistance (e.g., public telephone companies, social welfare and unemployment offices). However, even largely production-oriented companies can be partially service oriented; IBM, for example, is just one of many firms that provide after-sale service to customers who purchase the technology the firm manufactures.

To include the service sector in a definition of *innovation*, Kimberly (1981, p. 86) suggested a broad definition encompassing the ongoing economic shift from a mostly secondary to a more service-oriented economy and defined *innovation* as "any program, product or technique which represents a significant departure from the state of the art of management at the time it first appears and which affects the nature, location, quality, or quantity of information that is available in the decision-making process." This definition, however, does not really take into account the existence of computer or information systems, being too specific to be concerned with innovation in parts of the organizational system other than management. Like the other definitions that we have looked at so far, this one is inadequate for our purposes.

Because we are concerned with specific matters in this text, it is necessary that we have a specific working definition of *innovation*. The glimpses into previous innovation research we have already had provide the raw material from which we shall forge this working definition. For the purposes of this text then *innovation* should be understood as a process, or the product of a process, that is the result of the efforts or activities of an individual, group, and/or organizational system, which represents a departure from the previous state and may facilitate more effective resource allocation.

There are, theoretically, two kinds of innovation: product and process. Most of the earlier work in innovation has concentrated on distinguishing between these two types. Abernathy and Utterback (1978, p. 40) defined *product innovation* as radical change with product characteristics in flux. New products will result in the reorientation of corporate goals and/or production facilities. As we noted earlier, the creation and evaluation of new products and/or services are commonly departmentalized as research and development. Process innovation, on the other hand, theoretically involves incremental change, with improvements leading to a rigid, efficient production system specifically designed to produce a standardized product. The production of goods or services is typically departmentalized as design or product engineering.

More recent work in the area, however, has illustrated that the above distinctions between *product* and *process* may not always be representative of reality. The introduction of telebanking demonstrates that product innovations are not always radical and that process innovations are not always incremental. Telebanking was an incremental product innovation (24-hour banking increases customer convenience); however, it represented a radical process innovation by affecting banking operations (mechanizing and automating some tasks that formerly were done by bank tellers). In the early 1980s American banks started to offer high-interest-bearing checking accounts, a product innovation that has been possible only since regulatory change. This product innovation can hardly be described as radical, because banks have been offering checking accounts (albeit non-interest bearing) to their customers for quite some time (e.g., Nord & Tucker, 1987, chap. 3). On the other hand, when Credit Suisse, one of the major banks in Switzerland, introduced telebanking, it led to radical product *and* process innovation. The firm's product (banking services) changed dramatically as customers were then able to obtain up-to-the-minute information on foreign exchange rates, precious metal prices, interest rates, and other services from their home computer terminal via a videotext system at any time of the day or night. Visiting a branch office or interacting directly with an employee were often no longer necessary. Additionally, process innovation was introduced by improvements in the processing of customer orders and inquiries, thus improving efficiency and reducing transaction costs for the firm.

Another definition of *process* and *product innovation* is given by Rosegger (1986, p. 15), who distinguishes process innovations from product innovations by stating that the former involves "all those changes affecting the methods whereby outputs are produced, whereas the latter are aimed at the outputs themselves." He points out that in industrial practice, however, one rarely finds a new product that has not required some changes and adaptation in the process technology.

Distinguishing between process and product innovation may not be that important after all, because, where one type of innovation is found, we are likely to find the other. It is not always obvious at which point a process innovation becomes a product innovation (or vice versa). This may be a good thing; a binary distinction does not capture the reality of innovation and change, and process and product innovation should, therefore, be seen as representing a continuum rather than a dichotomy, with both types of innovation considered essential for an accurate description of overall organizational innovation and change. Because this book focuses primarily on how organizations adapt and integrate technological innovations in general, we will not bother to distinguish between *product* and *process innovation* in the future.

Technological Innovation

Now that we have a fairly clear idea about what *innovation* means in the context of this book, we can be more specific and explain what is meant by *technological innovation*. This is basically a matter of qualifying the previous definition: *technological innovation* should be understood to be a technology-based process, or the product of such a process, that is the result of the efforts or activities of an individual, group, and/or organizational system,

The Nature of Innovation

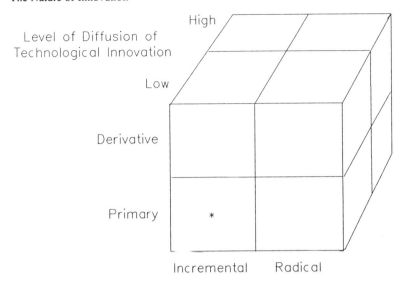

Figure 1.1. Classifying Technological Innovation
*To help in reading this figure, for example, an organization in the quadrant labelled * would adopt primary innovation incrementally, and it would choose innovations with a low level of dispersion.

which represents a departure from the previous state and may facilitate more effective resource allocation. Because the two definitions are so close, and because we are primarily concerned with technological innovation, the two terms will be considered synonymous for the rest of this book. As Figure 1.1 illustrates, a technological innovation can be classed according to type, diffusion, and relationship to its users. The taxonomy of technological innovation is discussed below.

Primary and Derivative Innovation

Although the line between *primary* and *derivative* innovation is not always clear, the basic difference between the two is the same as the difference between primary and applied research. *Primary research* is what leads to primary innovation and is based on work evolving out of the testing of theoretical relationships. The invention of plastic was a primary innovation, but its use led to derivative innovations that became the lifeblood of whole industries, such as, for example, the medical instrument industry. Primary innovations will eventually "trickle down" to reach *applied research*, which is responsible for the creation of derivative innovations — everything from plastic mop handles to microcomputers. The marketplace is the most important factor in determining the kind of derivative innovations that applied research will create. If there had not been a market for plastic handles or microcomputers, applied research would never have made them as common as they now are.

Our interest in the *acquisition of innovation and its accompanying organizational change* means that we are primarily interested in *derivative* innovation in this book. As an open system using input (raw materials and other resources) and transforming it into output (products and/or services), any firm needs to scan its customer base to determine what innovation may be necessary to remain competitive (Miller & Friesen, 1986). Most technological innovations in the area of computerization have been derivative, ignited by customer demand or by other environmental factors such as government regulations. However, because the line between primary and derivative innovation often blurs, we will not — though derivative innovation is the focus of our study — bar ourselves from investigating the grey area between the two types.

Innovation Diffusion

Diffusion describes the degree to which an innovation has become integrated into an economy. Rosegger (1986, p. 187) determined a list of factors (grouped into five major categories) found to affect the rate at which a technology diffuses. The first of these factors is the *origin of the innovation*. Innovations may originate from either end of a business relationship, and although manufacturers have an economic interest in seeing that their new products are used as extensively as possible, there is nothing to prevent a customer firm from making an innovation that it wants its suppliers to use. A second factor is *effects on other inputs*. Our definition of *innovation* states that it should result in more effective resource allocation, and innovations are frequently classified along labour-, capital-, materials-, or energy-saving dimensions. If more effective resource allocation becomes a demand of the market as the result of some other factor influencing the situation, diffusion rates will naturally be affected. Because of the high energy costs facing Swiss manufacturers, for example, there has been a rapid diffusion of a number of energy-conserving techniques largely ignored before the 1975 oil crisis. In another case, Swiss service industries (e.g., hospitality, banking, insurance)

The Nature of Innovation

faced with high labour costs, have adopted computerization in production processes and offices to an extent that is well beyond that of many other countries and that offsets the negative impact of these costs.

The third factor affecting diffusion is *the relationship of the innovation to the existing production structure.* Innovations that affect the total process of production will be more carefully scrutinized than innovations that only affect a part of the process. Banking machines have quickly become commonplace in North America and Europe, but completely computerized banking, which offers the customer access to his or her account via computer terminal, and which permits transactions such as the purchasing and/or selling of shares without the necessity of talking to a branch representative, is far less common. Banking machines only affect branch operations significantly by reducing labour costs, but some human labour is still necessary to make an "automated" teller feasible. Completely computerized banking, however, would affect a significant part of the actual production process itself.

The way the innovation is introduced into the production structure will also affect its relationship to that structure. An innovation that is introduced in an evolutionary pattern and is used by properly trained personnel is more likely to be a successful innovation and is more likely to have a high diffusion rate.

Change in the innovation is a fourth factor affecting its diffusion rate. Innovations undergo transformations during the actual process of diffusion, and the refinement of an innovation may suddenly make it useful to another industry. The innovations of NASA's space program will ultimately spread into civil industry through such a process. Expectations of rapid change, however, may actually *limit* an innovation's spread. The diffusion of the 16-bit and 32-bit chip technology used in some microcomputers was slowed (in 1987-1988) by the expectations of corporate users torn between betting on this system and waiting to see if the new IBM-PC OS/2 would become the industry standard that its predecessor was. Because end-users were unsure of the future compatibility of 16 and 32-bit technology with other machines and software, and were also unsure of alternative technologies, the acquisition of new systems was deferred by many decision makers, and the diffusion rates of all of the concerned technologies were reduced.

The last factor affecting diffusion is *complementaries among innovations.* Often, industry can only reap the complete benefits of one innovation and its acquisition if complementary innovations are adopted simultaneously. As we have just seen, end-users were reluctant to adopt innovative new computers because the innovative software that would determine their effectiveness was as yet unavailable. It is, therefore, not uncommon to find a series of innovations offered as a package. The "just-in-time" techniques used in manufacturing, which require changes in the design of assembly lines as well as in warehousing and purchasing processes, are often adopted as such a package. The greater the complexity of the package, however, the longer decision makers will hesitate before adopting it, and the slower the rate of diffusion.

Radical and Incremental Technological Innovation

The terms *primary* and *derivative* serve to delineate specific *kinds* of innovations and are constants in that a given innovation always occupies a specific place in the primary/derivative continuum (though we may argue about the actual location of that place). Yet, as Figure 1.1 shows, technological innovation can also be described as being either *radical* or *incremental*. These terms serve to indicate the relationship between an innovation and its end-user, and this relationship is necessarily a variable. An innovation that causes a radical change for the way one corporation does business may cause only small change for another.

A great deal of the recent research into technological innovation has been concerned with these two dimensions (see Miller & Friesen, 1986; Nord & Tucker, 1987). For our purposes, *incremental innovation* can be defined as the introduction of technology new to the organization that is very similar to its present technology and that requires little change in the processes of the organization to be used effectively. *Radical innovation*, on the other hand, is the introduction of technology that is very different from that which the organization has previously used, which, therefore, requires significant changes in processes if it is to be used effectively. Radical innovation requires personnel to adjust their behaviour and to acquire new skills to a degree far beyond that of incremental innovation.

Nord and Tucker (1987) showed that radical innovation interrupts the status quo. Other things being equal, the more changes there are in information, reward, and power structures, the more radical we must consider the technological innovation to be. Incremental innovation maintains the status quo; it may result in a situation in which the customer places his or her orders by computer terminal rather than by mail or by phone, eliminating some of the paperwork and shortening the time between order and delivery, but it leaves the business relationship essentially the same.

Given the effect of radical innovations on the status quo, we might expect them to be adopted only by individual firms, but entire industries do occasionally adopt radical innovations. Toyota is in the process of developing a new car engine that will allow it to meet U.S. air pollution standards *without* the commonplace catalytic converter. The implementation of this process will represent a radical innovation for Toyota, which anticipates radical changes in the production process for engines and in the skills required of the workers on the production line. What this forecasts for the entire industry might well be described as an automotive revolution, because other manufacturers may be forced to follow suit to remain competitive. Nonetheless, other considerations may make rapid change difficult, as we shall see in later chapters.

We can discuss radical and incremental innovation as being separate theoretically, but the real world, which defies attempts to make an absolute binary distinction between primary and derivative innovation, is equally defiant about allowing an absolute distinction between radical and incremental technological innovation. Once again, the distinctions we make between the two should be understood to be rules of thumb rather than absolutes, and this should be kept in mind when we speak of distinguishing between an organization that embraces radical innovation and one that favours incremental innovation.

Technological Innovation and the Firm: Some Illustrative Examples

Innovations allow firms and industries to respond to a variety of market changes, from rising energy prices to increased labour costs. For example, now that free trade has become a reality, Canadian brewers are in the process of closing some breweries and expanding others in their attempts to compete with American brewers, but if they wish to remain competitive they will have to upgrade their technology to American standards. Innovations allow industries to respond to nonmarket influences as well; in Europe, where pressure from environmental groups and parties such as the Greens is mounting, the acquisition of innovative technologies will likely be used to meet the guidelines set by more stringent pollution laws. We can make such educated guesses about how technological innovation might be adopted in the future because we have a wealth of examples demonstrating the effects of adopting technological innovation in the past. We shall examine some of these next, but given that we are concerned primarily with the acquisition of innovations, and thus with derivative technologies, we will not concern ourselves with organization-originating innovation, an interesting topic covered extensively elsewhere (e.g., Nord & Tucker, 1987; Pennings & Buitendam, 1987).

In the airline industry, information systems technology is used extensively to handle seat management and reservations. American Airlines could probably be considered the pioneer in this field, having initiated the Reservisor system in 1952. In 1962, this system was replaced with one based on a software package called ACP (Airline Control Program), which was later changed to SABRE (Semi-Automated Business Reservation Environment). This system has revolutionized the airline industry's way of selling tickets, administering seats, and managing its fleets by allowing it to plan schedules according to seats sold before the actual flight. Because the system's design was permitted by innovative primary work in the computer sciences, and because it represented a technological innovation made in response to the market (see Figure 1.1), it is clear that this innovation was derivative. It is also clear that it was incremental, as it was adopted in stages over the course of decades and is still revised periodically. In the United States, at least, the diffusion of the innovation has been quite extensive, given that about one-third of all airline bookings in that country are currently made on SABRE. Similar systems in Europe have not been nearly as successful, as we shall see in Chapter 3.

The overnight letter courier industry, which developed rapidly during the 1980s, owes its very existence to recent innovations. When Federal Express started in this business, it required information technology and sophisticated sorting equipment to guarantee next-day delivery. Later, a computer-based tracking system enabled the company to keep tabs on every single piece of mail, giving it the capability of keeping customers up to date on the exact location of their package at any given time. In the early stages, radical innovation was necessary to assure on-time delivery of a swiftly growing volume of overnight packages and letters. Later services, however, such as using facsimile systems to deliver letters on the same business day, built on the same technology used to transport overnight letters (with the possible exception of the fax machines themselves) and required only incremental change. Because overnight letter services are now available in most cities across North America and Europe, we can safely say that the diffusion is wide.

The BTX (*Bildschirmtext*) system is an interactive videotext offered by the German post office to subscribers, both organizations and individuals. Firms use BTX to communicate with customers on the system—another example of a radical innovation. Mail-order houses, car rental and travel agencies, government offices and manufacturers make use of the system, and

The Nature of Innovation

the post office theoretically earns money from both these agencies and their customers. Unfortunately, the system is running up a huge deficit because diffusion so far has been lower than expected. It may be that the innovation is currently too radical and that it may be more successful when the general public becomes more accustomed to computer technology.

Summary

Innovation is the sap that flows in the organizational tree, and the effective management of technological innovation is what makes an organization grow and flourish. When this sap is no longer flowing, the organizational tree becomes so much deadwood. This book studies the flow of this lifeblood within different organizations, each of which has its own subsystems — its own individual network of roots, branches, and leaves. Towards this end, this chapter accomplishes several things, the most important of which is establishing the definition of *technological innovation* that will be used for the course of this study. *Product* and *process* and *primary* and *derivative* have been introduced as types of innovation, and arguments explaining why this text will use a continuum for distinguishing between the former and the latter pair were given. *Diffusion* was discussed as an important dimension of technological innovation, and the difference between *radical* and *incremental* innovation was demonstrated using historical examples.

This done, our next step is to present the theoretical framework from which all further discussion must stem. Chapter 2 presents a model that combines the contingency and cross-cultural approaches to technology acquisition and the technology-induced change that attends it. By uniting these approaches, we are able to achieve a much clearer perspective of the multifaceted situation with which any organization adopting innovation is faced.

2. An Integrated Model of the Contingency-Culture Approach

The first chapter introduced the topic of technology acquisition and technology-induced organizational change, a subject that has gained substantial interest in recent years. The organization must adjust itself to new technology, but the technology must also have a certain fit with the firm's organizational structure, processes, values, and beliefs. The comparative analysis of organizations has been influenced by two major schools of thought: contingency and cultural theories.

Much of the research conducted with contingency theory as its base has attempted to develop theoretical principles that apply to organizations in general. Contingency theory states that, for an organization (or its subunits) to be effective, there has to be goodness of fit between its structure and environment, thus focusing on the *objective* perspective using a *macro level of analysis*. This macro focus attempts to increase our understanding of how environmental determinism and organizational choice may affect an organization's pursuit and accomplishment of its goals. Employees, managers and owners are constrained by the environment to adopt certain structural designs that affect the integration of new technology into production processes. Although research on this topic has abated somewhat in the past few years, the current literature continues to include reports of new theoretical and empirical developments, and several summaries and/or critiques of contingency based research have appeared (e.g., Pennings, 1987; Schoonhoven, 1981).

Another area of interest has been cross-cultural research on organizational behaviour. This area of inquiry concerns the systematic study of the behaviour and experience of organizational participants in different cultures. Cross-cultural approaches to management concentrate on *subjective* as well as *objective* dimensions, using a *micro* and a *macro level of analysis*. Cultural values, norms, beliefs, and modes of action as well as individual attitudes are studied. Although culture is an important concept in the social sciences, it has been defined in so many ways that no consensus has emerged and no systematic paradigms have been developed. Two independent developments have recently occurred, however, that may greatly advance our understanding of the role of culture and cultural differences in work organization, technology acquisition and technology-induced organizational change: the internationalization of business and of the social sciences.

In this chapter, we'll attempt to merge these two approaches and study technology and its introduction into organizations by developing a model that includes both the contingency and the cross-cultural perspectives. Each of the approaches is shown to help increase our understanding of the processes that occur when new technology is being introduced into an organizational setting. This chapter outlines the theoretical framework used in this book.

The Contingency Approach

The contingency proposition is complex because it hypothesizes a conditional association of *two or more* independent variables with a dependent outcome and directly subjects this to an empirical test. For example, the level of diffusion of a new technology might be expected to interact with internal labour market determinism to affect the organizational adaptation processes in a certain way. Because contingency theory tends to encourage classification of some variables as "independent" and others as "dependent," we might quite naturally assume that the relationship between these variables and their outcome is purely unidirectional, and that this relationship could best be examined by causal analysis (Schreyögg, 1980, 1982). However, functional analysis, which assumes that all variables are capable of influencing all others, proves to be more useful in practice than unidirectional analysis. Variables such as the organization's levels of internal labour market determinism (ILMD) and choice in strategic human resource management (SHRM) (see Chapters 3 and 4) are neither completely independent nor completely dependent, because their ability to influence and their tendency to be influenced are in a constant state of flux.

We can demonstrate this interrelationship between variables by looking at the relationship between an organization, Johnson & Johnson in this case, and its stakeholders (Mitroff, 1983). In isolated cases in the United States in 1983, bottles of Tylenol pills (an over-the-counter painkiller) were poisoned, making the product very dangerous in the public's view. Because the public is the largest group of stakeholders in any organization, their perceptions directly influenced the manufacturers to withdraw the product from all shelves across the United States. The organization in turn was able to influence the public; by doing *more* than simply complying with their demands and by introducing tamperproof packaging when the product was later restored to the shelves, Johnson & Johnson was able to win back consumer trust and goodwill. Today Tylenol is once again the leading over-the-counter painkiller in North America. The most effective contingency-based studies use a number of variables and examine their covariation statistically. Pennings (1987), for example, did not consider the environment to be a constant variable but assumed that both environment and structure exist in a multiple, dual-contingent relationship opposite indicators of organizational effectiveness. His study tried to test some approximation of choice, an approach suggested but not often attempted in the literature (e.g., Hrebiniak & Joyce, 1985). The results showed that high-performing units remained within their scope of choice (measured by such variables as market share and number of competitors), as presumed to exist under conditions of high congruence (fit) and effectiveness.

To illustrate the above, let us take a look at Wardair, a Canadian airline taken over in 1989 by Canadian Airlines International. Wardair held a relatively small market share for scheduled long-haul flights within Canada and competed against several major players (e.g., Canadian Airlines International and Air Canada). Its hierarchic structure was rather flat compared with its competitors, which reduced its overhead. Moreover, in response to market demand, it offered its customers a generous frequent flyer program in an attempt to attain high congruence (that is, fit between market demands and internal strengths equals quick response to frequent flyers' needs) and effectiveness (e.g., high loading factor and revenue per passenger mile). Although the company sometimes reacted to market demands, it also influenced the market by anticipating customer needs (offering better in-flight service at lower full-economy fares), thereby forcing its competition to adjust. Unfortunately for Wardair, competitors adjusted so well that the smaller company had a huge deficit in 1988, and its founder was forced to approach the holding company of Canadian Airlines International with a merger/takeover offer.

Although the contingency approach is very useful to the student of technology acquisition and technology-induced organizational change, it is flawed in that it tends to ignore the human element in the adaptation equation (e.g., Pennings, 1987). For instance, ILMD (rules and regulations pertaining to human resources such as staffing and promotion eligibility, see also Chapter 4) and the firm's choice in SHRM affect how it can respond to environmental constraints and opportunities. Certain human resource strategies, policies, and systems (see Part II) attract and help retain a certain type of employee in the firm, thereby creating a culture with specific values and beliefs. This culture will affect how the firm responds to new innovations and technological opportunities. Thus assessing such contingency factors as ILMD and choice in SHRM is a necessity, because they may either reinforce current values and belief systems or lead to desirable modifications (e.g., acceptance of new technology).

Figure 2.1 uses the factors that have been identified by earlier research as being important in organizational studies (e.g., Drazin & Van de Ven, 1985; Schreyögg, 1980) to expand the contingency approach into the internal labour market and strategic human resource management domain (Gattiker, 1988a). The model assumes that two types of interaction are present—*multiplicative* and *maximizing* (Schoonhoven, 1981). Multiplicative interaction presumes that effectiveness is high when high levels of both environmental and SHRM choice are present, and that it is low when either level of choice is low. Maximizing interaction supposes that there is an ideal value for levels of environmental and SHRM choice that will produce optimal effectiveness. We can enlarge this principle by stating that there should be a range within which the effectiveness of the firm can be maximized (Pennings, 1987). Hence, contingency theory would hold that such maximizing results in a relatively limited range of choice.

Part II of this book expands the contingency approach into the internal labour market and strategic human resource management domains (Gattiker, 1988a) and concentrates on the effects of systems components on technology acquisition and technology-induced organizational change. Part III then studies the individual employee — possibly the most important participant in the firm (March & Simon, 1958, chap. 4) — and the effects of individual perceptions and attitudes upon this process. These perceptions and attitudes are most easily examined using a *cultural approach*. Although the contingency and cultural approaches to technology acquisition and organizational adaptation have traditionally been perceived as being distinct and separate fields of investigation, it is nonetheless true that our understanding of the processes involved in technology acquisition and organizational adaptation will not be improved by ignoring either approach.

Cultural Approach

The introduction of computer-based office information technology is usually driven by a desire for financial savings and productivity increases,

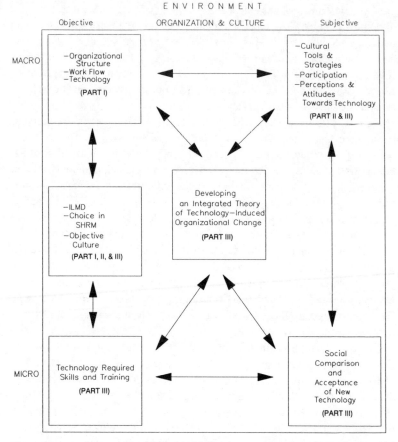

Figure 2.1. Technology Acquisition and Technology-Induced Organizational Adaptation as Seen Within Its Cultural and Environmental Context Using Both the Objective-Subjective and the Macro-Micro Dimensions

but employers seldom consider the potentially negative effects technology may have upon employees and their perceptions of the quality of work life[1] (Gattiker, 1984). This approach is, to some extent, forced on management, because the effects of innovation on personnel are often uncertain at the time of a technology's introduction into the workplace (Kahn, 1981). As a consequence, the company seeking the competitive edge implicit in new technology must also cope with the "growing pains" common to innovators.

Perceptions of career success, of the quality of work life, and of any new technology should be very important to the organization intending to adopt the technology (Dierkes & von Thienen, 1984) — especially because these perceptions may affect satisfaction, work roles, absenteeism, and organizational commitment (see Gattiker, 1988b; Locke, 1976; Mowday, Porter, & Steers, 1982, chap. 3). *Organizational commitment*, as defined by Porter, Steers, Mowday, and Boulian (1974), has three major components: (a) a strong belief in, and acceptance of, the organization's goals; (b) a willingness to exert considerable effort on behalf of the organization; and (c) a definite desire to maintain organizational membership. When perceptions of the effects of a new technology lead employees to become dissatisfied, their commitment to the organization will suffer.

James and Tetrick (1986) tested several alternative models of causal relations between job perceptions and satisfaction, and their results suggest that job satisfaction occurs *after* job perceptions in the causal order. This supports the claim made by Lazarus (1982, 1984), who suggested that precognitive causal models, in which perceptions occur after attitudes, are *theoretically* untenable. The core issue in Lazarus's model was that "cognitive processes are always crucial in the elicitation of an emotion" because emotional responses in humans are elicited "by a complex cognitive appraisal of the significance of events for one's well-being" (Lazarus, 1982, pp. 1024, 1019). James and Tetrick's study also suggested that precognitive models are *empirically* untenable, leading them to identify a need to develop and test alternatives to their postcognitive-nonrecursive model. My work (Gattiker, 1988b) has shown that the postcognitive model is applicable in the office-automation domain by testing a postcognitive-recursive model and finding that perceptions of computer-mediated work affected organizational commitment, role conflict, and role ambiguity.

It would be a mistake to examine individual perceptions of innovation without making an attempt to understand the antecedents of these perceptions, and this is where the cultural approach becomes relevant. As earlier research seems to suggest, perceptions are affected by work environments, group structures, and other organizational characteristics; as well, external environmental factors may influence workers' perceptions of technology-triggered organizational change. For instance, a performance appraisal system that assesses desirable responses to technology-induced organizational change, and a bonus system rewarding such newly acquired behaviours, will reinforce positive attitudes in employees and in turn help the change process. If the cultural context of the organization is understood, many of these environmental and organizational characteristics are explicable.

As we will see in Part II, cultural contexts account for the differences in high school education in Japan, the United States, and West Germany and also for the differences between the on-the-job training needs of organizations in each of these countries. It would be even more correct to qualify the preceding statement by saying that cultural contexts cause the *differences in perceptions* that account for the dissimilarities between these countries. Unfortunately, because most of the research on technological innovation and organizational adaptation has been domestically focused — with some exceptions (e.g., Gattiker & Nelligan, 1988; Maurice, Sorge, & Warner, 1980) — research into perceptions and their effects tends to address important issues by examining subjects in one cultural setting without specifically describing and analyzing the cultural context or addressing cross-cultural issues. Although organizational research has been justly criticized for doing so (Sydow, 1985, 1987), it has not shown any change towards a more comprehensive approach.

Because cultural contexts differ, a model for discussing technology acquisition and technology-induced organizational change based on contingency theory alone may not always provide a completely realistic picture of the state of any given organization in a specific situation. The contingency model advances the "culture-free" hypothesis (Hickson, Hinings, McMillan, & Schwitter, 1974; Hickson & McMillan, 1981; Hickson, McMillan, Azumi, & Horvath, 1979) which argues that certain contextual variables (e.g., choice in SHRM) and dimensions of organizational structure (e.g., determinism of the ILM) are similar across very different societies (see Chapter 10 for an extensive discussion of this issue). However, were we to look, for example, at two large organizations in West Germany and Japan both experiencing low ILMD and high choice in SHRM, we could reasonably assume that, although their similar situations would cause them to act similarly in some ways, their different cultural contexts would cause them to act differently in others (Pettigrew, 1985, chap. 5; Swidler, 1986; Wilpert, 1986).

This would suggest that, if the cultural and contingency approaches were integrated, we would be able to develop models for technology acquisition and organizational adaptation that are more representative of reality. Combined, the two approaches would offer a far more comprehensive model for studying and managing technology acquisition. Some contingency research has already begun this proces: Miller and Droege (1986), for example, reported that the relationship between perceptions and structure was usually greatest in samples of small, young organizations, particularly because in such firms the chief executive wields great authority, and his or her personal desire for achievement (based on his or her personal perceptions) might be as influential upon structure as structure is on it.

Level and Focus of Anaysis

Most organizational research assumes that a person's behaviour in the firm is rational or at least quasi-rational. Use of the individual as the unit of analysis is based on the supposition that he or she will act with conscious purpose. However, bounded rationality theory revokes this assumption. The essence of this concept is the premise that there is a limit to the amount of information the human mind can process when a rational decision becomes necessary (Simon, 1957, p. 158). Human information processing carries with it four major consequences. First, the individual's perception of information is not comprehensive but selective. Second, people tend to process information sequentially because they can only integrate small portions simultaneously. Third, one's memory depends upon operations that simplify judgmental tasks and reduce intellectual effort. Finally, and most importantly, overall human memory capacity is severely limited (Simon, 1984).

People attach more weight and importance to information and data that they consider to be causally related to a target outcome (Hogarth, 1980, pp. 42-43). Bounded rationality presents significant implications for a model evaluating technology acquisition and technology-induced organizational change because the assessment is governed mainly by the manager's, employee's, or stakeholder's (e.g., shareholder, customer, and supplier) different frames of reference.

Based on the above, it is clear that neither contingency theory nor cultural theory alone can really account for the technology acquisition and organizational adaptation phenomena. Katz, Kahn, and Adams (1980) have pointed out that the individual and the organization must be used as units of analysis to increase our understanding of technology-induced organizational change. Thus both a macro approach, using organizational variables such as differentiation of structure, coordination, organization of work, and hierarchic control and communication systems, and a micro approach, using individual variables such as perceptions, attitudes, beliefs, and needs, are necessary.

The use of both individual (micro) and organizational (macro) units of analysis must, however, be complemented by using objective (e.g., profit, productivity, costs) and subjective (e.g., individual feelings, attitudes, perceptions, and beliefs) measures. While objectivity may result in goal oriented or purposeful decisions, subjective factors may impede people's ability to think rationally. Thus the focus of our analysis must span the objective-subjective dimension of inquiry.

The approach to technology acquisition and technology-induced organizational change proposed below uses different levels of analysis to develop paradigms that will increase our knowledge about organizations and their potential for successful use of advanced technology and innovations. This part outlines a model that uses four measure categories, based on two dimensions: (a) level of analysis (i.e., individual and organizational) and (b) subjectivity-objectivity.

A Contingency-Culture Model

An organization's dependence on various resources requires that it interact with those who control these resources. Generally speaking, the resource-dependence model proposed by Pfeffer and Salancik (1978, chap. 10) has been used to examine external participants (stakeholders such as stockholders, bankers, and customers) who control certain resources that are critical to the success of the firm and who, thereby, are able to influence the organization's actions and behaviours. In the Tylenol case mentioned earlier, the buying market possessed a critical resource (money) which became less available to Johnson & Johnson after its Tylenol pills were poisoned. It might be nice to imagine that the company pulled the product from the shelves solely in the interest of its buyers, however, it is perhaps more realistic to say that this action had more to do with public outcry that the company could not ignore if it still wished the public to buy its products and thus provide the company with the resources it needed.

In Part II of the book, the crucial external participant is generally subsumed under the environment, represented by a group like a union. However, there are many different critical stakeholders. One of the most important of these is the employee, who, although in part dependent upon the organization for employment, is in turn depended upon by the firm for his or her effective participation in the transformation process by which the firm turns a variety of raw resources into some sort of product (Scott, 1981, pp. 109-110).

It is no secret that technology in general, and computers in particular, have revolutionized the world of work during the past two decades. Technological advances in data processing, telecommunication, decision support systems, and office automation continue to affect virtually every aspect of organizational and work life. Paradoxically, we are in a rather poor position to assess the impact of these technologies on organizations and the people who work with them. For example, although computers have long been used by employees as part of an information system for decision and control purposes, and as a production and work tool, few attempts have been made to develop methodological approaches to the study of this phenomenon.

Industry in developed countries is increasingly moving towards the service sector (e.g., banking, insurance, and the hospitality industry), and administrative work is expanding rapidly. Organizations depend upon the end-user's (employee's) effectiveness in processing and communicating information using new, advanced technology. In most cases, mechanization of certain tasks has occurred but rarely has total automation taken place.[2] Thus end-user computing serves as an important example for assessing the model's usefulness in developing research paradigms for the study of technology acquisition and technology-induced organizational change.

Micro and Macro-Level Analysis

Pfeffer (1982, p. 15) argued that possibly the most fruitful approach to organizational research is to match the unit of analysis to the level at which the theoretical processes underlying the phenomenon being studied are operating. In the case of Figure 2.1, it is obvious that the level of analysis is both micro (individual) and macro (system), because the theoretical processes underlying the phenomenon operate at both levels.

Figure 2.1 proposes a dual approach to technology acquisition and technology-induced organizational change. Thus the book begins with a macro perspective (organization) and focuses mainly on objective and national factors influencing the organization's culture (labour laws, unions, and education). The remainder of the book deals with micro determinants (individuals and groups) and subjective culture (e.g., values, norms, and perceptions) to illustrate further how they may affect and be influenced by technology-induced organizational change.

Both contingency and cultural approaches initially use a macro perspective to look at the organizational system and then use a micro perspective to examine either subsystems (in the case of the contingency approach) or the individual (in the case of the cultural approach). Functional analysis suggests that the elements examined by the contingency approach will influence those that are the concern of the cultural approach and vice versa, which implies that a match between all of these elements needs to be accomplished. If, therefore, matching between the environment, SHRM, and culture can be accomplished, the multiplicative factor will be a firm's increased effectiveness in organizational adaptation of technological innovation.

Subjectivity and Objectivity Dimension

Figure 2.1 describes technology acquisition and technology-induced organizational change within the *cultural and organizational environment* and suggests that a multivariable approach to technology acquisition and technology-induced organizational change is needed. Why? We have often viewed organizational and individual variables as being of such a different order that appropriate relationships could seldom be established (Katz, Kahn, & Adams, 1980, p. 7). People believe in the purity of levels and the undesirability of mixing psychological and structural measures. This error arises from a failure to conceive of firms as systems that are nested within, and linked to, larger systems (e.g., markets). Furthermore, we must remember that firms contain smaller subsystems (see Chapters 5-9) that are, in turn, linked to them. Although the rational approach to organizations suggests that objective measures alone are needed to explain technology acquisition and organizational adaptation processes, we have already established that, given the human proclivity for irrational behaviour, a subjective perspective is not only helpful but necessary to comprehend actions occurring within an organization (e.g., March & Simon, 1958; Pfeffer, 1982, chap. 2).

Summary

As we have seen, there are two dimensions or continua that we can use to study technology acquisition and technology-induced organizational change: contingency theory and cross-cultural management theory. In the following sections, each part of the model will be discussed more extensively.

The remainder of Part I outlines in more detail how environmental determinism (e.g., market regulations) and strategic choice can affect organizational culture, technology acquisition policies, structure, and processes. Part II then concentrates on explaining how the interrelationship between a firm's ILMD and choice in SHRM can affect its human resource policies, strategies, and systems in such areas as recruiting, socialization, performance appraisal, and training.

In Part III we shift our analysis to the subjective side of the model to explain cultural factors, attitudes, and beliefs held by the employee about the organization and about technology-induced change in particular. We will examine the interrelation of Part II and Part III, and explain why human resource systems, values, and beliefs (or simply the organizational and national culture) may affect each other to determine a firm's potential for successful technology acquisition. The final chapters of this part concentrate on social comparison processes and employee acceptance of, or resistance to, new technology and computers in particular.

The Contingency-Culture Approach

I will reiterate at this point that, although the book focuses in some parts upon *end-user* technology, the latter term is rather all encompassing. Thus, in this book's context, the end-user is any individual who works with a variety of end-user technologies (e.g., information systems, robots, computer-aided design, computer-aided instruction, computer-based manufacturing and air-traffic control systems) that use computer components such as memory chips.

One major focus of this book is the individual. No matter what type of technology is used in the firm, the individual stands at the centre, and his or her effectiveness in using this expensive technology ultimately determines an organization's success or failure. The individual is at the core of the technology management challenge in the conceptual framework presented here.

Our next step is to discuss and explain how different types of organizations deal with technological innovation, its acquisition, and organizational adaptation. Because market conditions, the actions of competitors, government regulations, and a firm's own strategy affect its general approach to technology acquisition, we need to consider the interrelationship of environmental determinism and level of organizational choice, two continuous dimensions that have usually been perceived as being dichotomies. This will be the focus of Chapter 3.

Notes

1. Quality of work life encompasses, but is not necessarily limited to, an individual's subjective evaluation of his or her work and career situation. As such, job features affected by the organization's adaptation to new technology — career success, stress, organizational commitment, and job characteristics — are part of the overall construct of a person's quality of work life (see also Chapter 11).

2. Automation differs from mechanization in that it involves handing over the control of certain processes to computer-based technology (e.g., Doswell, 1983, chap. 1).

3. Strategic Choice and Environmental Determinism

An organization may adopt technological innovation in the form of manufacturing technology, computerized production, or management information systems. Whether or not it actually does so is often dependent upon a variety of environmental constraints, such as the degree of choice the firm may have and the degree of environmental determinism limiting this choice. Different internal and external influences may cause new technology to be introduced and adapted by the organization in different ways; acquisition may be a process reflecting choice and selection by the firm or one in which it is a necessary reaction to peremptory environmental forces.

As outlined in Chapter 1, to smooth the transition process from "old" to "new" technology, the firm must obtain a certain fit between the new technology and internal structures. Incompatibility requires the firm to adapt. This chapter develops a typology of technology acquisition and technology-induced organizational change based upon these constraints, discussing organizational choice and environmental determinism as two independent variables on separate continua. These two variables are considered using the dimensions outlined in Chapter 1 and Figure 1.1 (e.g., level of diffusion, type of innovation, and process of introduction).

Introduction of New Technology and Organizational Adaptation: Choice Versus Determinism

During the past few years, organizational research has begun to explore the impact of technology acquisition upon organizations (e.g., Pava, 1983; Roznowski & Hulin, 1985). *Technology acquisition* in this context refers to the organization's decision to introduce or adopt a technology/innovation. The *proactive* organization freely chooses to introduce untried and innovative technology early on (radical innovation), and this is likely to give it a competitive edge in the marketplace. The more radical the innovation, however, the greater the risk the firm takes when deciding about acquisition. At this early-adopter stage, goals may not be easily quantified, thus making a financial assessment of the potential risks taken by the firm rather difficult (see O'Reilly, 1983).

The *reactive* organization, on the other hand, only adopts technology in reaction to a threat that is the result of environmental changes, such as a competitor's acquisition of new technology. When a firm is reactive, it may have to adopt innovation with a high diffusion rate. Technology managers can thus base their decisions on quantifiable projections of risks and objectives (e.g., O'Reilly, 1983) because other firms have gone before them and "ironed out the bugs" of adopting the technology.

There are advantages and disadvantages of adopting innovations before they have become widely dispersed. An organization that adopts an innovation early is like a person who buys fabric and is then faced with the difficulty of fashioning from the cloth a custom-tailored suit that precisely fits his or her measurements and tastes. At an early stage of development, an innovation may also be custom-fitted to suit a firm's particular needs and circumstances, and from this tailoring process an organization gains valuable knowledge of the adaptation process, which can be applied to future technology acquisition. In contrast, a person who buys clothing off the rack or a firm that buys technology off the shelf reduce the financial risks because they can see exactly what they are purchasing. But at the same time, they forfeit an intimate knowledge of the way the garment or system is put together.

Such knowledge is extremely helpful for successful technology adaptation within an organization. The term currently used in the literature is *organizational adaptation*; it is used in a number of ways, and depending upon whose definition we choose, it may refer to both *proactive* and *reactive* change (Miles & Snow, 1978) or (more specifically) to the process of reacting to *environmental* changes (Astley & Van de Ven, 1983). The usage in this book aligns more with the former, allowing for both reactive and proactive adaptation and change. *Organizational adaptation* in this context refers to the organization's attempt to gradually integrate technology into its production processes, structure, and information processing systems. *Reactive adaptation* occurs when a firm decides to start with the necessary internal changes shortly before the innovation is being introduced, as when a firm informs its employees about impending changes due to the introduction of new office technology a week or less before the computer arrives at a person's desk. *Proactive adaptation* ensues when a firm tries to anticipate and assess needs for internal changes and introduces these in a gradual manner before the innovation is adopted, perhaps even allowing employees to participate in the decision to choose a particular technology.

To illustrate proactive and reactive adaptation and their interrelation with technology acquisition, we can look at the introduction of the Airbus 320 at Swissair and Air Inter of France. Swissair told its pilots that advanced technology could reduce the cockpit crew required for the A320 from three to two members. They also explained the potential of the A320 to help in strengthening the company's market position (e.g., via operating and labour cost savings). During the acquisition phase, they discussed adaptation processes required to make effective use of the new technology. Only after the pilots' union gave its consent for the necessary technology-induced changes and reductions in the cockpit did management make the decision to adopt the new technology and order the plane. The necessary training efforts for pilot crews were initiated long before the plane was delivered. In contrast, Air Inter ordered the planes before discussing their impact upon job security and job responsibility with the pilots, thereby separating the acquisition decision from the adaptation phase. Pilots went on strike several times during 1988 and the planes could not be used effectively because crews refused to fly them, using the argument that they were not safe. Faced with these inconsistencies between new technology and internal interests (e.g., labour savings versus the pilot's concerns for job security), Air Inter reacted by offering its pilots guaranteed job security and increased compensation to finally convince them to fly the planes, a very costly *reactive* approach.

To date, much of the literature exploring the relationship of technology acquisition and adaptation to organizational factors such as structure, corporate strategy and environmental determinism has been characterized by an external focus. This may be an unfortunate development, insofar as technology acquisition and organizational adaptation seem most easily understood by comparing *both* internal and external contexts. Internal contexts will become the focus of our study in later chapters, but at present we shall appraise the externally focused literature to provide a basis for later comparison. We will also explore implicit relationships with several conceptual traditions in organization theory.

Environmental Determinism and Strategic Management

Salancik and Pfeffer (1978, chap. 1), citing numerous examples, have shown that the environmental factors constraining corporate behaviour have long been a concern of organizational research. Research has indicated, for example, that a factor such as high interest rates can hamper a firm's effective adaptation of new production or other technology (e.g., Arnould, 1985). More recent research specifically assesses the impact of environmental constraints on corporate strategy (e.g., Miller & Friesen, 1986). In this context, *strategy* is used in the sense of *strategic management*, the decision-making process that focuses the capabilities of the organization upon the opportunities and threats it faces in its environment (Rowe, Mason & Dickel, 1982; see also the section "Strategic Human Resource Management" in Chapter 5 for further discussion of this topic). Unfortunately, however, the existing literature characteristically seems to assume that a clear binary distinction between the continua of determinism and choice in strategy captures the reality of organizational behaviour and change. As popular and intuitively pleasing as such a distinction may be, it is misleading, tending to overemphasize the explanatory power of one continuum while deemphasizing the explanatory power of the other. It directs attention away from the fact that both are essential to an accurate understanding of organizational behaviour and change. Because the organization, consciously or not, essentially interprets the two continua as interrelated, and because this interpretation affects the decision making of the organization, it is important that we see them as interrelated.

Hambrick (1981) proposes the two constructs of *environment* and *strategy* for explaining natural selection and the adaptive models that describe power structures within organizations. His data reveal the interactive effects of environment and strategy upon power patterns and other organizational processes. Hambrick's model finds support in other research; Miller and Friesen (1980), for example, propose that certain types of organizational transitions over time (e.g., stagnation and consolidation) are linked to environmental, organizational, and strategy variables. More recently, Lengnick-Hall (1986) has proposed several strategy archetypes to achieve more effective computerized batch production, assessing not only structural changes but organizational design as well. A firm's characteristics and environment quite clearly affect the strategy it pursues and the organizational change it undergoes.

Processes of Introducing New Technology

Just as certain environmental conditions can be expected to suppress innovative behaviour, other conditions seem to have a tendency to promote it. There are historical precedents to which we can refer: Goldstone (1987), comparing the Stuart kingdom and the Ming and Ottoman empires—three governments at crossroads in the seventeenth century — discovered three factors that distinguished England from China and Turkey in this period that may have led to Britain's greater industrial innovativeness: (a) a *belief* that innovation was possible; (b) *social pressure* that either rewarded innovation or punished innovators who were not productive; and (c) a *political climate* that encouraged (or at least tolerated) rather than suppressed innovation. In the modern world, a political climate healthy for innovation can likewise encourage innovation in various ways. Governments can subsidize firms or allow them to depreciate investments made for adopting new technological innovations on a faster time schedule than otherwise possible, and educational policies might facilitate the acquisition of advanced degrees by keeping tuition fees low, thereby providing well-trained people to work with the new technology. Of course, just because it is possible for a government to promote innovation through these methods does not mean that it will do so, and the tonic value of the present political climate for innovation might be debated. It would be difficult, however, to argue that our technology-oriented society does *not* believe innovation is possible, and equally difficult to argue that innovativeness goes unrewarded. So it would seem that Goldstone's three characteristics of an innovative society are as easy to find in modern innovative societies as in postmedieval ones.

Although Goldstone was concerned with the attitudes of societies in general towards innovation, we are necessarily more concerned with those of individual organizations within a larger society. These attitudes are, of course, affected by other factors, such as the effects of market conditions and competitors' actions or the promise of tangible benefits from innovations, such as cost reductions in materials and labour or productivity increases. The interrelationship of these factors will determine whether the introduction of an innovation into the organization will be *evolutionary* or *discontinuous*. As it should be understood here, an *evolutionary* introduction of innovation is step by step, whereas a *discontinuous* introduction would occur if a firm attempted to implement new and unfamiliar technology within a rather short time frame. An organization may have little control over the way it introduces innovations, and the actions of competitors may force a firm to make a quantum leap in technology to remain competitive, by introducing technology discontinuously that might preferably be introduced evolutionarily. This will be especially likely if the foreseeable benefits of the innovation are great. Nonetheless, a *proactive* acquisition and adaptation strategy is more likely to lead to an evolutionary introduction of new technology. To illustrate these two approaches, we return to the American Airlines computerized reservation system discussed in Chapter 1. SABRE evolved over decades into today's multiuse information system, offering a variety of services for travel-related activities (e.g., ticketing, car rentals, hotel reservations and charge accounts). European airlines, in contrast, have only recently decided to cooperate to form a computerized reservation system similar to SABRE, and they are locked in an intense battle to dominate the air transport industry's computerized reservation markets. Although some airlines are participating in Amadeus (e.g., Air France, Iberia, Lufthansa, and SAS), others are part of Galileo (e.g., British Airways, KLM, Swissair, and United Airlines). Neither system, however, had the time to introduce the necessary technology in the preferred evolutionary manner, because both chose 1989 as the target date for offering a full range of services. Discontinuous innovation was forced upon them primarily because of the threat of SABRE and other North American competitors trying to establish themselves in the European markets.

The SABRE system has a distinct advantage over its European counterparts in that (in theory, at least) most of its bugs have already been worked out in the course of the long evolutionary process through which it has come into existence. Amadeus and Galileo, in contrast, are relatively untried systems that still have to prove their merit. The point has been made, and cannot be overemphasized, that technological innovation is essentially irrational. Adopting any new technology, even if it is similar to one already used, cannot be considered an instrumentally rational perfecting of the system. Regardless of what some choose to believe, the consequences of introducing new technology can only be pure conjecture. There is thus an element of risk in the acquisition of an innovation, and the risk is higher the more rapid the implementation. Even attempts to slowly integrate technology occasionally fail, however, and if an innovation fails to accomplish management's goals, a carefully thought-out pattern of evolutionary implementation may have to be scrapped in favour of a jury-rigged discontinuous plan to help the firm cut its losses or catch up with its competition. Discontinuous implementation may be made possible by acquiring the technology and knowledge from another organization. The disadvantage of this approach is, however, that experience gained during the evolutionary implementation phase will probably be of little benefit when discontinuous innovation becomes necessary. The latter may require totally different skills, thereby increasing the costs for the firm to remain up to date with its technology.

Typology of Technology Acquisition and Organizational Adaptation

Despite their close relationship, organizational choice and environmental determinism can be considered independent variables in the technology

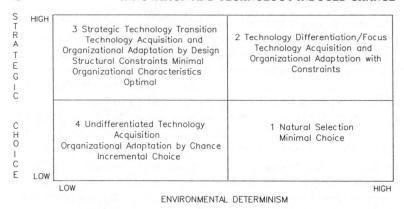

Figure 3.1. Relationship of Strategic Choice, Environmental Determinism in Technology Acquisition, and Organizational Adaptation by the Firm

acquisition and organizational adaptation process. Although individuals and organizations do not live within an entirely deterministic world, and although they can make choices in decision-making circumstances and construct, eliminate, or redefine the objective features of the environment to expand their own decision-making capabilities, the effects of the environment remain important nonetheless. Certain characteristics of industries and national economies are intractable, immune to the control of individuals and their companies. Figure 3.1 examines these two interacting variables of choice and determinism, which can be represented on axes ranging from low to high. Each axis denotes variance in levels of assertiveness and potential to influence others (Gattiker, 1988a). The quadrants help to define the domain and scope of power in the relationship between organization and environment and the relative vulnerability of each in an interactive setting (Pfeffer & Salancik, 1978). In a large organization, it is even possible that individual divisions within the firm can be placed in different quadrants.

Quadrant I basically shows the conditions or assumptions underlying the *population ecology/natural selection* approach to technology acquisition and organizational adaptation, which argues that organizations exert virtually no control over exogenous factors. Acquisition is in part determined from outside the firm, as the environment selects the firm and allows only those with appropriate variations and products to remain profitable. Under these conditions, firms must either acquire technology according to external determinism and make the necessary organizational adaptations or cease to exist (Williamson, 1985).

Quadrant I firms are usually found in markets that are highly regulated. Generally, these organizations exercise little control because a regulator in the environment dictates the price of their product while the demand for it can be fairly elastic. Canadian rail companies find this situation familiar. Prices for shipping certain commodities (e.g., wheat) are set by the government, while demand for shipping capacity can be elastic (e.g., drought or bumper crop might affect overall shipping volume). Differentiating products to command higher prices and profits is nearly impossible. Railways that fall behind and do not integrate new technology into their transformation process risk finding their costs for labour and technology rapidly rising. These environmentally imposed constraints give the organization little choice, which makes it difficult to establish a lasting competitive edge.

At the opposite extreme are firms existing under the more munificent conditions of Quadrant III, marked by high levels of organizational and strategic choice. Technology acquisition and organizational adaptation are, therefore, proactive, so that autonomy and control by the firm are the rule rather than the exception. The organization in Quadrant III confronts a pluralistic environment in which movement between market niches is not severely constrained by exit or entry barriers. According to Snow and Hrebiniak (1980), companies in this quadrant neither lack scarce resources nor experience many political constraints, and are, therefore, able to define their domains and the exogenous conditions under which they desire to compete. Within Quadrant III, organizations acquire technology and adapt to it by design. Introducing technological innovation is easier due to the benign environment, which encourages the emergence of "prospectors" who are to the business world everything that their name connotes — individuals or companies looking for the next "gold rush" in the form of an innovative new product that will mean overnight success (Kimberly, 1987).

Many of the large consulting firms in Europe or North America illustrate organizations in this situation. Exit or entry barriers do not constrain their markets to any great degree and they experience a certain autonomy and control over their destinies. These organizations have the opportunity to purposely influence the environment and decide under which conditions they desire to compete in a national market.

In Quadrant II, both strategic choice and environmental determinism are high, defining a turbulent context for the acquisition of new systems (Miller & Friesen, 1986). Under these conditions, there are certain clear indigenous factors that affect decision making. Nonetheless, the firm enjoys choice despite the preeminent nature of external forces and constraints. Typical cases here would include firms in market niches in which certain rules or immutable environmental conditions compel certain outcomes or behaviours but allow leeway and choice in others. Technology acquisition and organizational adaptation is less restricted because companies and industries produce a variety of products and services.

Financial institutions such as large banks are often found in such a situation, with regulations limiting some of their products and services while not affecting others. Capital requirements and legal constraints are relatively severe, but the organization still has a certain autonomy regarding what type of services it may offer its customers and under what conditions. The tobacco industry has thrived by finding market niches and affecting its potential consumer base through extensive advertising, marketing, and lobbying, despite government-required health warnings placed in advertisements and on packages — another example of an industry surviving with only limited control over its environment.

Quadrant IV represents a relatively stable situation for a firm, yet firms in this group tend to lack strategic choice despite a paucity of external constraints. Changes in technology seem to occur almost by chance, because these firms do not seem to exhibit coherent strategy to take advantage of fortuitous market and product conditions. One explanation for this puzzling tendency of Quadrant IV organizations may simply be that their internal strengths and competencies are inappropriate for strategic technology acquisition and, therefore, they do not take advantage of the possibilities open to them. Assuming that management wants to move the firm away from this condition, it means that executives must first develop different skills needed to take advantage of internal conditions.

As this analysis suggests, the acquisition and adaptation process is dynamic; over time, an organization's position may shift as a result of its own strategic choices and changes in its operating environment. Although organizations that are involved in similar business activities are often ruled by similar environmental situations, they are not always found in the same quadrant, as we shall see in later examples. A technology acquisition and technology-induced organizational change model has to integrate such possible alignments and change processes to represent the adaptation stages realistically.

Technology Acquisition and Organizational Adaptation Issues

Table 3.1 demonstrates how organizations in different quadrants approach some of the main issues and problems associated with introducing new

TABLE 3.1: Four Organizational Types: Approaches to Technology Acquisition and Organizational Adaptation

Variable	Quadrant I High Determinism, Low Choice	Quadrant II High Determinism, High Choice	Quadrant III Low Determinism, High Choice	Quadrant IV Low Determinism, Low Choice
Number of firm-initiated technology changes	Few	Medium-high	High	Few
Stage strategies for adopting technology	Defender	Differentiation focus: analyzer	Differentiation focus: prospector	Reactor
Process of introduction	Discontinuous	Evolutionary	Evolutionary/Discontinuous	Discontinuous
Organizational adaptation strategy	Reactive	Proactive or reactive	Natural progression and proactive	Reactive
Type of innovation	Radical	Incremental/radical	Radical productivity and profitability driven	Incremental
Diffusion of technology	High	Medium-high	Low	High

SOURCE: Based in part on Gattiker (1988a).

technology, thus adapting their organizational structure and processes to assure a fit.

Firm-Initiated Technology Changes

Because levels of choice vary widely across the quadrants, it follows that the number and type of strategic options vary correspondingly. There are few viable options for firms in either Quadrants I or IV, because environmental determinism limits change in the former and internal factors inhibit decision making in the latter. The number of changes is highest in Quadrant III, where environmental determinism is relatively low and organizational choice is high. In Quadrant II, the number of changes depends upon environmental and internal organizational factors.

A comparison of companies providing similar services in different quadrants might be helpful at this point. US Sprint, an American telephone company in Quadrant III, enjoys a deregulated market and its structure and organizational characteristics allow for rapid change. Its acquisition of recent technological advancements (including the installation of a fibre-optic network) has led to a number of additional technology changes. In contrast, the Post, Telephone and Telegraph (PTT) of Switzerland, which is in Quadrant IV, is part of a highly regulated market, which limits its ability to introduce new technology and make the subsequent organizational adaptations. This organization was required to get approval from the federal regulatory commission before introducing videotex (an interactive information system used via phone lines that requires a modem, computer screen, and alphanumeric keyboard), which slowed down the acquisition of that innovation substantially. However, although the Swiss PTT's situation is not encouraging for innovation, it must be noted that deregulation and increased innovation in telegraph and telecommunication markets would, in this case, actually *reduce* the annual profits of this federal agency. British Telecom, representative of Quadrant II, experiences a medium to high level of technological change (new services and so on) while the German *Bundespost* (the German PTT) in Quadrant I experiences high determinism, because worker participation in the decision-making process is legally required before introduction of new technology, and low choice, because it is under the jurisdiction of a special federal ministry, the *Bundespost Ministerium*.

Strategies for Adopting Technology

Firm-initiated changes as well as other items in Table 3.1 can be better understood by looking at the generic strategies used by firms as defined by several researchers (e.g., Porter, 1980; Snow & Hrebiniak, 1980). Porter developed three such strategies: *differentiation* strategy, which aims at creating a product or service that is somehow different; *cost leadership* strategy, which aims at making the organization the lowest-cost producer in an industry; and *focuser* strategy, which caters to a circumscribed and specialized segment of the market such as a certain kind of customer, a limited geographic market, and/or a narrow range of products. Additionally, Miller and Friesen (1986) have proposed the *reactor* strategy, which is used to respond to the actions of competitors, and the *defender* strategy, in which the firm tries to defend its market share, typically by lowering prices or by dumping.

Although a firm in Quadrant I, for example, may use defender strategies when it comes to technology and especially research and development (e.g., Hitt, Ireland, & Goryunov, 1988), it will not necessarily use a defensive product strategy; instead, it is likely to use a mixture of strategies (Miller & Friesen, 1986). Nevertheless, an organization can usually be assigned to one group according to the single strategy it tends to prefer. For example, risk-taking, creativity, and innovation are clearly more consistent with the ideal conditions facing the prospector in Quadrant III than with the conditions in, say, Quadrant I. The unstable organization that uses the reactor strategy as opposed to initiating change, characterized by no real agreement on outcomes and lacking a clear focus when it comes to technological acquisition, would appear most likely to find a home in Quadrant IV.

Within a large corporation, different divisions may experience different levels of determinism and make use of a variety of strategies characteristic of different quadrants. To illustrate this, we can look at Sony's experiences in the videocassette recorder (VCR) industry. A few years back, Sony was a prospector, introducing VCR machines that used their own Beta-system technology when there was not yet a definite market for them, which placed the firm in Quadrant III. Unfortunately, although Beta systems indisputably offered better replay and colour quality than rival systems using the VHS format, the latter were invariably cheaper. When the smoke cleared after a vicious marketing war, it became clear that VHS was victorious in Europe and North America. The environment demanded a specific kind of product and would settle for nothing different. Sony was forced into becoming a defender — the characteristic position of Quadrant I firms. Finally, in January 1988, the company admitted defeat and became a reactor by deciding to sell VHS machines. In 1988, up to 20% of the VCRs the company sold were VHS. The reaction is characteristic of Quadrant IV, yet, given the company's past history of innovation, it is likely that it will eventually be able to return to the more desirable strategies characteristic of its place in Quadrant III.

Introduction Processes

In Quadrant I, firms are usually defenders. Because they do not introduce innovations until forced to do so by environmental pressures, they usually have no time for the evolutionary approach and introduce innovation in a discontinuous fashion. Examples of innovations introduced in this way are the forementioned Amadeus and Galileo systems. Quadrant II firms, on the other hand, prefer an evolutionary approach and have the freedom to use it (see Perry & Sandholtz, 1988). By being analytical and trying to differentiate themselves from other organizations, these firms will attempt to advance continuously by introducing new innovations. The earlier described SABRE airline reservation system was the product of such an attempt, and it demonstrates what American Airlines managed to achieve by moving itself out of Quadrant I.

A less fortunate example is that of American steel producers, who have often been criticized for reacting to environmental change without a clear, strategic, long-term focus. U.S. Steel, in attempting to move out of Quadrant I, should have realigned its strength and tried to differentiate to find its market niche in the steel business (the logical move for a Quadrant I organization). Instead, the company transferred its resources out of the steel business by purchasing Marathon Oil, a business that, as its name suggests, had nothing to do with the traditional interests of U.S. Steel. Although this could be considered an extremely radical attempt to differentiate the firm's product from that of its competitors, no synergy effect or vertical integration could truly be said to have justified such a step. We might well ask why the management of U.S. Steel believed this would increase the firm's profits and why it did not instead invest the resources paid for Marathon Oil (which some shareholders felt were exorbitant) into its traditional line of business and thus try to improve its situation—but it is unlikely that we will ever have an answer. We can guess that internal factors (e.g., union and human resource policies) may have prevented the firm from exhibiting a coherent strategy to take advantage of fortuitous market and product conditions. Ironically, it seems that the company was prevented from moving to a more preferable quadrant by the very short-sightedness that such a move hoped to eliminate.

We might expect a similar approach towards the acquisition of technology from Quadrant III firms, given their freedom to make choices, but environmental factors such as prospecting and the progress of competitors are wild cards that often force the discontinuous introduction of technology in this quadrant. Automated banking is a fairly unregulated industry and is thus representative of this quadrant. When certain North American banks started to evolutionarily introduce automated tellers that allowed customers to withdraw cash at any of their branches' machines; competitors, taken by surprise, were forced to introduce similar services discontinuously to keep up. Then, when some banks, such as First Interstate, began to offer customers the option of using banking machines to withdraw cash anywhere in the West from *any* machine that was part of their First Interstate network, others once again had to scramble to keep up, by joining national clearing networks like Exchange and Cirrus that allowed their customers to use, for a fee, other banks' machines that were part of the network.

In Quadrant IV, the process of introduction is discontinuous, because acquisition and organizational adaptation happen by chance owing to the lack of evolutionary strategies (see Kamm, 1987). An example might be the Union Bank of Switzerland (an exception to the rule of large banks belonging in Quadrant II), which in the early 1980s tried to computerize most of its operations in a short time to react to its competitors. The introduction of the technology was discontinuous and initially not very successful, although this situation was eventually corrected. Today the Union Bank of Switzerland is again in Quadrant II, having a clear technology strategy allowing for an evolutionary introduction of further innovations.

Organizational Adaptation Strategy

Quadrant I firms are most likely to respond reactively to technological innovation because they avoid making changes to their structures and processes, as needed for smooth integration of new technology, for as long as possible. When mail sorting technology was first developed, private courier services (Quadrant III; e.g., Federal Express and Emery) implemented a proactive organizational adaptation strategy prior to technology acquisition, which allowed them to make effective use of the technology shortly after its introduction. This gave them a competitive edge over the U.S. post office, which used a reactive organizational adaptation strategy and is still struggling to make productive use of the technology.

A Quadrant IV firm is usually forced to be reactive when it comes to technology-induced organizational change. For instance, Canadian railway companies have adopted technology for freight trains that makes the conductor in the caboose dispensable. Unfortunately, the railway did not pave the way for this transition by negotiating with the union concerning potential layoffs and transfers prior to adopting this technology. Instead, organizational adaptation was reactive, contributing to costly strikes by railworkers in 1988.

Innovation Types

Quadrant I firms are typically faced with radical technological innovations because they tend to avoid change until they are absolutely forced into it, usually by a threat to their existence. Canada Post, the Canadian government postal service, is a good example of an organization in this situation, cowed by one of the most powerful unions in Canada and adopting innovations only when competition from private courier services becomes too great.

Firms in Quadrant II may introduce either radical or incremental technological innovation, depending upon the company's particular differentiation strategy and the market (Miller & Friesen, 1986). Quadrant III firms introduce radical technological innovation with the hope of making major productivity improvements, and it is only through such risk-taking that prospecting can be sustained. Credit Suisse, for example, introduced the innovation of telebanking in 1986, which at the time represented radical innovation; however, the additional services that the system was providing by 1988, such as listing share and bond prices, represent only incremental innovations. The business relationship is still essentially the same, although some paperwork has been eliminated and certain additional services can be provided more easily and quickly.

In Quadrant IV, firms will usually choose to introduce incremental innovation because they have not yet developed the strengths needed to adopt radical technological innovation. Some banking institutions in this quadrant have added automated bank tellers incrementally, albeit rather tardily. This was done without a clear technology strategy but was necessary to remain somewhat competitive as this service has essentially become standard in the banking industry.

Technology Diffusion

In Quadrant I, organizations adopt innovations only after their use by others in the industry has become widespread. An example would be the U.S. steel industry, which, in the early 1980s, introduced technology already in use by its Japanese competitors. Quadrant II firms will probably adopt new innovations with a medium to high diffusion in the marketplace, using analyzer strategy to guide the acquisition of new technologies only after they see the benefits competitors gain (Gold, 1983). After American Airlines' computerized reservation system SABRE became a success, most of its major competitors, such as United Airlines, either introduced their own systems or attempted takeovers of airlines with such systems, which was the case when Texas Airlines took over Eastern Airlines.

Quadrant III firms will adapt to technology that is not yet widely used in the field; again, Credit Suisse is exemplary of firms in this quadrant. Quadrant IV includes firms that can be characterized as reactors and will most likely adopt technology characterized by a substantial diffusion. The automated bank teller machines mentioned in the previous section represent the type of innovation these firms might adopt.

Summary

As we can see by looking at the typology formulated here, the two variables of environmental determinism and organizational choice determine four different types of organizations, each of which typically possesses distinctive decision-making processes and attitudes towards the acquisition of technological innovation and each of which can be characterized by a governing principle. The principle governing organizations in Quadrant I of our model is *natural selection*, and the organization controlled by this has a minimal degree of choice in adopting new technologies; it must either adopt and make the necessary organizational changes or cease to exist, usually by going bankrupt, losing its market share, or being taken over. Quadrant II organizations are governed by *differentiation* and possess a relatively high degree of choice while remaining subject to a degree of environmental determinism high enough to constrain the acquisition of technology. Organizations in Quadrant III, characterized by *strategic choice,* enjoy maximum choice and adopt new technology by design, making proactive changes. And those in Quadrant IV are governed by *undifferentiation*, feature incremental choice, and adopt new technology by chance. These four types influence the number and forms of strategic options open to organizations and determine the type of innovations that the organization chooses to introduce — a decision that the diffusion of the innovation in the marketplace will affect differently in each case.

Given what we have learned in this chapter, the ideal location for an organization is Quadrant III. Positioning a firm in this quadrant allows management to differentiate its products from those of its competitors, lower production costs, and increase the profitability and well-being of the organization (see Miller & Friesen, 1986). As a result, the organization is in a position to offer its employees a certain degree of job security and a stable income based on its own wealth and success.

Firms in Quadrant IV should make a serious effort to restructure themselves and move toward Quadrant III, with which they already share a similar freedom from environmental determinism. The natural objective of firms in Quadrant I should be to move either to Quadrant II — which would require an internal restructuring similar to that which a firm moving from Quadrant IV to Quadrant II makes — or to Quadrant III — although this would, of course, be quite difficult. Moving toward Quadrant III would allow an organization to introduce more radical innovations that are not yet highly diffused in the market and also permit it to introduce new technology evolutionarily rather than discontinuously. Unfortunately, such a move is risky, and although it may prove to be of great benefit to the organization, it may also have disastrous consequences if it is not carefully planned, as the example of U.S. Steel illustrated.

This chapter has discussed the interactive nature of the organization-environment relationship and its effect upon technology acquisition and technology-induced organizational change. It has shown that acquisition of new technology also necessitates organizational adaptation. A final important implication of the present analysis is that simple models, relying on the conceptual construction of mutually exclusive, competing explanations of cause and effect, may not be sufficient to capture the intricacies of technology management. Technology-induced organizational change is the subject of inquiry in such diverse disciplines as economics, industrial engineering, management, organizational behaviour, and sociology, all of which emphasize different and often conflicting assumptions, foci, and explanations of why firms adopt technology and what its impact will be. As this chapter suggests, what is needed is a greater emphasis on integration rather than differentiation of views. The theoretical and conceptual barriers between the literatures of these disciplines must be reduced, permitting an eclectic approach when seeking explanations for technology adoption and technology-induced organizational change.

4. Determinism and Organizational Characteristics

Like any creature, the organization is the product of an evolutionary process that enables it to compete within its environment. When the environment changes, only those organisms that can change with it will survive. The organization, however, is of that rare species that can adapt itself much faster than any other; rather than waiting for evolution to take its natural course, it can *decide for itself* the way it will adapt. Although such forced evolution occasionally proves disastrous, it often enables organizations to dominate their environments. In most instances, the decision by management to adopt a new technology or innovation requires organizational adaptation to obtain a smooth fit. Hence, management must change or adjust organizational structure and processes in such a way as to make the most effective use of the new technology being introduced. Proactive adaptation by the organization most certainly smooths the acquisition process of new technology, while reactive adaptation may defer the potential benefits of the new technology.

In this chapter, we examine the four different types of organizations presented in the previous chapter, assessing their overall characteristics and the effects of these upon technological innovation. We also examine how the different types of organizations are distinguished by more specific internal characteristics, such as communication structures and organization of work, and how these affect the adoption of new technology and organizational adaptation.

Four assumptions guide our analysis. The first is that any organization will strive to maximize its transformation process. Whether a profit or nonprofit firm, a government agency, or a cooperative, an organization attempts to use input resources as effectively as possible. The second assumption gives the motivation for the first; it states that an organization conducts its activities in such a way as to maximize the benefits for its stakeholders (who include *everyone* with an interest in the company's activities — shareholders, employees, customers, members of the general public who are affected by the company's endeavours, and so forth). Organizations must be concerned with satisfying stakeholders because they are *open systems* (a topic discussed in greater detail below), and it is, therefore, important that they avoid becoming bureaucracies. In the private sector, *bureaucracy* describes any open system whose administrative and production processes have been slowed down and are extremely cumbersome (Meyer, Stevenson, & Webster, 1985). Bureaucratization, of course, makes it difficult for the firm to satisfy its stakeholders.

The third assumption is that there are no absolutes governing the interaction of organizations and their environments. A variety of strategies for responding to the environment are possible, depending upon the situation and the organization's preference. One organization, for example, may lay off workers to respond to lower sales, and another may first try to shorten the workweek for all its employees to reduce production and thereby avoid layoffs. Given the absence of a predetermined system, however, management will attempt to invent one, using innovation to increase its influence over the direction of the firm and employing numerous controls on a frequent basis to provide the organization with information about its performance when looking at different divisions/functions of the firm. The fourth and final assumption is that environmental and organizational determinism lead to certain preferences as to how the firm will be structured and organized. The different quadrants introduced in the last chapter are distinguished by different ideas about what forms of organization are appropriate for which purposes; for example, what form of organization provides the most effective means of managing technological innovation.

General Organizational Characteristics

All organizations are complex and different, so it may be helpful to begin by focusing on their basic similarities. As open systems, organizations are dependent upon a flow of resources and personnel from outside. To survive, a firm must induce its participants (e.g., employees, shareholders, and customers) to contribute their own resources and energy: Shareholders, for example, must be convinced to provide capital and/or invest time in the firm by sitting on its board. Though these participants commit themselves to the organization because they believe it will provide them with the best return on their investment, they also have outside interests that may not coincide with those of the organization. The open system must, therefore, be seen as a coalition of individuals with various, constantly shifting interests, which

Determinism and Organizational Characteristics

develops and agrees upon goals by *negotiation*. The structure and activities of this coalition will be influenced largely by the environment and by general characteristics such as *ownership, governance, distinctive competencies,* and *structure*. Some researchers have indicated that such internal characteristics may be more important to successful organizational adaptation than external ones when new technology is introduced (Young, Hougland & Shepard, 1981).

Ownership

Large organizations are commonly managed by individuals who are employees and who might also own a small percentage of company stock. That small percentage, however, can be quite important in determining managerial behaviour, especially (as some research has shown) in expense-related situations (Arnould, 1985). Bank managers' shares of ownership, for example, are usually limited, so banking executives tend to be expense focused, typically leading them to maximize pecuniary gains. This expense preference behaviour has created costs that have reduced the reported profitability of banks (Arnould, 1985). In contrast, high ownership by management will change the executives' approach to being more profit or future focused.

Executives with limited ownership, faced with the decision either to adopt innovation or to refrain from doing so, focus mainly on reducing costs. Based on his or her degree of opportunism and ownership, the individual tends to make decisions that maximize his or her benefits. Bounded rationality and limited information processing (see Chapter 2) force such managers to concentrate on measures that are quantifiable and that make the organization's most important stakeholders judge their performance favourably (e.g., O'Reilly, 1983). High ownership may allow managers to be less concerned with short-term profitability and cost gains and possibly more mindful of the long-term benefits of adopting a new technology.

Public organizations such as government agencies and utilities are not that different from private companies when it comes to considerations of ownership. Because executives of public organizations do not own any shares, they tend, like the managers with little or no ownership, to make decisions that maximize personal benefits. The decision-making process potentially differs insofar as public organizations may have objectives that cannot be evaluated by quantifying the problem, but the manager will nevertheless still try to satisfy the most important stakeholders.

Governance

The governance structures of most organizations are usually quite similar in their basic design. In general, profit organizations are controlled by owners who are also shareholders. Management is assisted by a board of directors, which may include voting and nonvoting members representing such diverse groups as employees, management, unions, customers, and suppliers. The different affiliations and goals of these board members can place serious constraints upon management's ability to make decisions, and the priorities of some board members and their constituencies may affect organizational adaptation efforts to smooth the process of adopting innovation (see Birkwald, 1973; Cameron & Whetten, 1983; Mitroff, 1983, p. 35). A rational manager will try to satisfy the most important members of the governance structure (see Larwood, Gutek, & Gattiker, 1984; Simon, 1984).

Public organizations such as school districts or universities have a slightly different governance structure from private firms. Nonetheless, these types of organizations are also governed by a board of trustees or governors who represent different stakeholders with various needs, objectives, and interests. Once again, these stakeholders have to be satisfied by management to sustain their continuous participation in the system, while added environmental constraints such as budget cuts may further limit an organization's ability to realize its goals.

Distinctive Competence.

Selznick (1957) first used the term *distinctive competence* to describe the character of an organization. It refers to those things that a company does especially well when compared with others within a similar environment (Andrews, 1971; Snow & Hrebiniak, 1980). In one research project, competence was measured by having managers rate each of 10 broad functions on a three-point scale that indicated areas of strong, weak, and average competence. The functions used were: (a) general management; (b) financial management; (c) marketing/selling; (d) market research; (e) product research and development; (f) engineering, basic and applied; (g) production; (h) distribution; (i) legal affairs; and (j) personnel. The authors reported that most managers indicated strong competence for their companies in *all* functions (Snow & Hrebiniak, 1980), coinciding with the results of Stevenson's (1976) study, which found that managers at the top level of their firms perceived more strengths than weaknesses in their organizations.

Most successful companies will have strengths in areas in addition to those mentioned above. These strengths might be in communication systems, organization of work, or organizational culture (Morgan, 1986, chapters 4, 5), areas that may add a great deal to the firm's competency in introducing innovation into the organizational production process. Distinctive competencies for public organizations may not differ significantly from those of private ones. If public managers also perceive their organizations to have more strengths than weaknesses, these organizations are likely to do as well for their participants as private companies.

Organizational Structure

The most visible part of life in any organization is the daily, weekly, and quarterly cycle of routines, procedures, reports, and forms. We usually understand *organizational structure* to be a configuration of these types of decision-making and information processing activities, activities that are characteristically enduring and persistent. Patterned regularity is the dominant feature of this configuration (Ranson, Hinings, & Greenwood, 1980).

Organizational structure can be described as being either *unitary* (U-form) or *multidivisional* (M-form). The U-form organizes the firm's activities along functional specialties such as sales and finance, while the M-form uses quasi-autonomous operating divisions, organized mainly along product and geographical areas (Williamson, 1985, pp. 133-136). Small firms generally find the U-form sufficient, and the M-form is more attractive to the larger firm (Williamson, 1985, p. 284).

U-form and organizational adaptation to innovation. U-form organizations usually require careful coordination of strategic direction and control for all departments. The care required sometimes means that the adoption of technological innovation and technology-induced organizational change is slower than it would be in a larger company using the M-form. Young, Hougland, and Shepard (1981) found that small U-form-organized banks were less innovative in competitive environments than large, M-form-organized ones in less competitive environments. However, we must remember that banking is a highly regulated industry in which small organizations are often forced to protect themselves from larger ones by offering better service at lower service charges or interest rates—these organizations cannot *afford* to be innovative. If the environment is *favourable* and less regulated, on the other hand, small- to medium-sized firms using this form will generally adapt to innovation more rapidly than large, U-form firms because their size facilitates the coordination of change. Centralized authority and a clear policy regarding technology that facilitates the rapid adoption of innovations and their integration into the production process may well give the small firm the advantage (see Dewar & Dutton, 1986; Ettlie, Bridges, & O'Keefe, 1984). At what point the U-form becomes less attractive than the M-form is, however, somewhat less than clear from the literature (e.g., Burton & Obel, 1984; Williamson, 1985). For a large organization, the U-form's central authority and the cumbersome procedures one has to go through before a new innovation can be approved and implemented may prove to be great disadvantages.

M-form and organizational adaptation to innovation. The M-form organization is extremely popular, having become widespread in the last 50 years throughout the United States and (more recently) Europe (Burton & Obel, 1980). The M-form, as Williamson (1985, p. 284) states, "combines the divisionalization concept with an internal control and strategic decision-making capability." Studies seem to suggest that reorganization to the M-form generally allows for higher financial returns than the U-form provides. The M-form, with its well-developed, general corporate office, solves the administrative control problem via an internal resource allocation system analogous to the function of external capital markets. Each division becomes a profit centre where divisional managers are rewarded based on divisional operating profitability. Operating responsibilities are dispersed among divisional managers. With its corporate staff freed from operational decisions, the top management group can focus its energies on the problems of overall strategic direction and entrepreneurial action. The M-form's ability to solve these two problems makes it more useful than the U-form to firms of large size.

Hoskisson and Galbraith (1985) studied several large American organizations attempting to change from the U-form to the M-form, some of which used an incremental approach and some of which used a radical approach. Time-series data were used to compare efforts and their outcomes. The results indicated that reorganization to the M-form generally produces a positive effect on performance and that incremental reorganizations had less effect on profitability than the radical transitions.

The M-form should encourage the introduction and adaptation of new products that represent incremental departures from existing process technology. Ettlie, Bridges, and O'Keefe (1984) found that organizational size correlated with decentralization and formalization, and concluded that the M-form may offset some of the less desirable effects of adopting innovation by enabling the firm to adapt to technologies more easily than with a U-form, assuming that the firm is large. Foster and Flynn (1984) described how individual divisions and departments in General Motors, an M-form type of organization, could adopt information technology and extend its use without being held back by the corporate office as would have been the case using the U-form.

Organizational Characteristics in Four Organizational Types

Although we have previously used the variables of environmental determinism and organizational choice to predict attitudes towards technological innovation, we may find it more enlightening to examine the relationship between these variables and specific organizational characteristics because these items more directly determine an organization's approach towards technological innovation. Table 4.1 does exactly this, using the four organizational types presented in the previous chapter and outlining their internal characteristics. Examples of representative organizations follow.

Quadrant I

An example of an organization in Quadrant I is the Fletcher Challenge conglomerate of New Zealand, which is a major force in the forest industry in Canada as well as in its home country, deriving more than half of its profits from forest- and paper industry-related activities. The company is publicly owned, which is to say that its shares and those of its subsidiaries are listed on several stock exchanges. Governance is as open to the influence of various stakeholder groups as it is to that of the shareholders; environmentalists, for example, affect its forest management strategy by lobbying against cutting trees in a certain area. The company uses the U-form structure (functional organization — production, sales, and so on), and its distinctive competencies include engineering and basic and applied research. It is also competent in legal affairs, being often under public scrutiny.

Public utilities and companies owned by the government such as Canadian Crown corporations may also belong in this category. Because the politicians controlling these corporations are elected and are quite vulnerable to the demands of their respective constituencies, such companies are often forced into making certain kinds of decisions (such as giving contracts to subcontractors located in politically strategic districts) and are thus severely limited in their choices. Again, stakeholders have a strong influence over management, although they lack formal authority.

Quadrant II

Organizations in Quadrant II are usually public and governed by a board of directors and various stakeholders. General Motors (GM) is a prime example. The company's board, made up of large investors, the house bank, top management, and other knowledgeable individuals, represents the formal governance structure of GM. The board assures that shareholders' interests

TABLE 4.1: Four Organizational Types: General Characteristics

Variable	Quadrant I High Determinism, Low Choice	Quadrant II High Determinism, High Choice	Quadrant III Low Determinism, High Choice	Quadrant IV Low Determinism, Low Choice
Choice	Minimum	Differentiated	Maximum	Incremental
Ownership	Public/government	Public	Public or private	Private, public, or government
Governance	Various stakeholder groups	Various stakeholders and the board	Owner(s) and board	Various stakeholders
Distinctive competence	Engineering, basic and applied research, legal affairs	General management, financial management, engineering, basic and applied research, production, and legal affairs	General management, financial management, marketing, product research and development, engineering, basic and applied research, production, and legal affairs	Financial management and basic research
Organizational structure	U-form	M-form	M-form	M-form or U-form

SOURCE: Based in part on Gattiker (1988a).

are served by the organization. However, the firm's customers represent a particularly powerful and important stakeholder group and can have a substantial effect upon the organization's policies and strategies. If, for example, consumer advocates declare a car unsafe, as happened with some types of J-cars around 1980, a serious threat to the organization's well-being can occur.

The type of structure GM uses is the M-form, which resulted from a radical reorganization in 1925 (Hoskisson & Galbraith, 1985), and has resulted in divisionalization of its operations. Such a structure, however, requires competence in both general and financial management. Production competence has improved continuously over the last decade, although it is weaker than that of Japanese competitors.

Quadrant III

An illustration of an organization in Quadrant III is Louis Vuitton-Moet Hennessy, a conglomerate that produces cognac, champagne, leather goods, and perfume. Though its stock is listed on the Paris stock exchange, and though governance structure is adhered to by a board, there are a few major stockholders (e.g., Vuitton family, 16%; the Moet & Hennessy family, 10%; and Guinness, 12%) that control the company and make policy. The board functions as an advisor to the majority shareholders more than anything else. The merger between Louis Vuitton and Moet Hennessy has led to divisionalization of the firm. The firm shows distinctive competence in various areas of management; its financial acumen is demonstrated by its complex but well-thought-out capital structure, and it is also strong in marketing and research. Making cognac, leather goods, and other products requires distinct competence in production and product research.

If a substantial portion of a company's stock is held by an individual or a small group of individuals, which is the case with Hoffman La Roche (a Swiss pharmaceutical firm with such products as Valium and Librium), management enjoys the privilege of making occasionally unpopular decisions, so long as they fit the long-term strategy. Roche's unsuccessful attempt to acquire Sterling Drug Inc. was not necessarily a financial masterstroke (it was extremely expensive), but a successful takeover of Sterling Drug Inc. would have suited the firm's long-term strategy of obtaining a stronger hold in the over-the-counter drug market in the United States. Management of a company with widely held capital may not enjoy such freedom. The little investor depending upon a regular and "high" dividend may resist the kind of behaviour exemplified by Roche management.

Quadrant IV

A firm in Quadrant IV may be privately, publicly, or even state controlled. Alberta Government Telephones (AGT) is such an organization, owned by the province but supposed to be managed like any other private firm. Even though its labour costs are very high and its average service (e.g., installing new residential phone lines, upgrading transmission using fibre optics) is deficient, changes are not in sight. Competency in marketing (e.g., customer services), finance (high debt ratio), and human resources (personnel assessment and planning) is not particularly high. In addition, politicians still have substantial influence upon the board and the overall governance of AGT, and the firm's structure is a U-form with some M-form components. Unless environmental constraints are decreased and competency in weak areas can be improved, it is unlikely that the firm can be moved to another quadrant that would offer more choice.

Ownership, governance, distinctive competencies, and structural types — these characteristics vary according to the levels of determinism and choice experienced in a given organization, which is to say that they vary according to the four different organizational types we have discussed here. Next, we discuss several internal characteristics that go in tandem with these types and more directly determine attitudes towards innovation.

Organizational Types and Internal Characteristics

Nature of Tasks

As we can see by looking at Table 4.2, the tasks a firm may face range from producing a limited line of products and services to producing a wide variety of innovative ones. The nature of an organization's production process will necessarily affect its willingness to adapt to innovation. An organization that has geared its entire production process to one standard product, as is typical of Quadrant I firms, would, of course, be unwilling to adopt innovations that would require the complete reconfiguration of the system. At the other extreme, an organization in Quadrant III, geared towards producing a wide variety of products, would find it much easier to implement new technology.

A Quadrant II organization, although faced with high levels of environmental determinism, is relatively free to make choices. One such organization is Mövenpick, a Swiss-based worldwide restaurant chain. Most restaurants

TABLE 4.2: Four Organizational Types: Internal Characteristics

Variable	Quadrant I High Determinism, Low Choice	Quadrant II High Determinism, High Choice	Quadrant III Low Determinism, High Choice	Quadrant IV Low Determinism, Low Choice
Choice	Minimum	Differentiated	Maximum	Incremental
Nature of task facing the firm	Efficient production of standard product	Efficient design, production, and marketing of new products and services	Wide range of innovative products and exploitation of rapid technological change	Limited range of products/services, modified by customers
Organization of work	Clearly defined jobs in hierarchic pattern	Jobs defined by interaction of job holders with others	Jobs are not clearly defined	Jobs defined according to functional and hierarchic pattern
Authority structure	Based on organizational structure; influence affected by seniority	Informal pattern of authority; based on influence	Authority and responsibility are vested in individuals; influence is based on expertise	Follows hierarchy: different situations may cause change
Decision-making and negotiating strategies	Centralized, extensive internal negotiating policies and strategies	Decentralized, strict negotiating rules and consideration of other participants' needs and feelings	Decentralized, flexible negotiating strategies and policies	Centralized; focus on resource and ideological negotiation
Communication system	Mainly vertical; based on rules and regulations	Guided by rules and conventions; junior staff encouraged to consult with management group	Free across all levels: central to organizational functioning	Free across all levels (though rarely used); primarily within work group

may be found in this quadrant because they have necessarily high levels of environmental determinism (e.g., health regulations, building specifications, and so forth). Mövenpick uses this freedom to adapt itself to the innovations that it requires to provide better services at competitive prices, enabling it to compete in the lucrative franchise market. If this organization lost such freedom of choice, it would be in Quadrant I and lack the ability to use innovations to provide unique culinary experiences and adapt to the demands of operating effectively in different parts of the world — it might produce one dish that it was extremely well known for and little else. If the company were in Quadrant III, on the other hand, and faced relatively low environmental determinism, it might spread all over the world and produce all varieties of foods, because it would not have to worry about the kinds of regulations that put a damper on this sort of business at present and that make prospecting difficult. Because it would be able to provide whatever kinds of products it wished, it would have the freedom to experiment with innovations, adopting what it needed whenever necessary. In Quadrant IV, however, the lack of choice would likely have prevented the company from ever taking the risk of becoming an international firm in the hospitality business, and the production of a limited range of products would discourage it from doing any more in the way of adopting innovation than "keeping up with the Joneses."

Organization of Work

From the earlier discussion of organizational structure, it follows that the organization of work will vary according to the typology, especially because — as Meyer and colleagues (1985, chap. 7) have noted — it is dependent upon the same combination of organizational choices and preferences that determine organizational structure that the typology reflects. The connections between the structure of a firm and the way its work is organized are impossible to ignore. For example, in a Quadrant III organization, jobs cannot be clearly defined because the organization is likely to be structured according to the M-form. An accountant in one subsidiary may do different tasks from an accountant in another, a manager in one division may have different responsibilities from one in another, and so on. The situation is similar in Quadrant II organizations because the M-form is the norm here as well, although the U-form, common in the other two quadrants and most common in Quadrant I — the least innovative quadrant — is more conducive to hierarchic patterns and jobs clearly defined according to job descriptions or handbooks. It is not difficult to draw a correlation between a certain freedom of activity and innovativeness; neither is it hard to make a connection between strictly defined jobs and strictly confined minds. Because innovation, by definition, involves breaking from traditional ways of doing things, the individual in Quadrants I or IV is unlikely to be found having anything to do with innovative activity, which could lead to charges of failing to carry out his or her job properly.

Authority Structures

The individual who holds a clearly defined position within a hierarchy is most likely to find clearly defined authority attached to that position, and this is supported by findings in the literature (e.g., Blau & Schoenherr, 1971, p. 121; Mansfield, 1973, p. 478; Scott, 1981, chap. 3). However, this authority is necessarily limited and falls within clear formal boundaries. Organizations in Quadrants I and IV will feature this sort of authority structure, and although it may be a bit more efficient than the looser structures based on influence and expertise that are the features of Quadrants II and III, this system discourages the adoption of innovation because the individual with the authority to decide whether or not an innovation should be implemented may not know much about the matter and is discouraged by the system from leaving the decision to those who do. In Canada Post, for example, where authority is dependent upon seniority rather than skill, new mail sorting technology would never have been adopted if the final decision fell to an incompetent who was "set in his ways" and was blind to the benefits of the technology. Though the structures used in Quadrants II and III are occasionally inefficient because no one's word is law in a situation in which the question of whose expertise is greater is debatable (in which case an appeal to higher authority might be necessary), the system is quite open to innovation because the opinions of the appropriate experts within the company are important. In a company like Mövenpick, for instance, a special week featuring the cuisine of one particular country would require substantial planning and organizing, and in such matters a cook may have more influence on the food and beverage manager than his or her job level would normally give. The openness of the communication system (which we will discuss below) also means that an appeal to higher authority is less of a problem than we might expect.

Decision-Making and Negotiating Strategies

Organizational effectiveness could be described as the fulfillment of the goals of the organization's dominant coalition of stakeholders. As we noted earlier, the fulfillment of goals is often dependent upon the successful resolution of conflicts between participants through negotiation. Table 4.2 indicates that different organizational types use different negotiating strategies to resolve conflict within and outside the organization. In keeping with the U-form structure that Quadrant I favours, negotiations are centralized and carried out according to strict rules. Decision making is, therefore, done by the coalition with the greatest amount of power (Pfeffer, 1982, pp. 74-80). Quadrants II and III, on the other hand, use the M-form structure and feature decentralized decision making. As we can see from the table, environmental conditions determine whether or not internal negotiating in these kinds of organizations will be flexible or guided by clear policies and strategies. In Quadrant IV, decision making is centralized, but in contrast to Quadrant I, a definite policy for resource and ideological negotiation exists. Mutual agreements, contracts, and position statements between participants are different according to the conditions under which they are formed.

How may decision-making and negotiating strategies affect adoption of technology and the necessary technology-induced organizational change? Centralized decision making can slow down the adoption process — even the modification of organizational structures has to be negotiated before any changes can be made. All this will make adoption and organizational implementation more time-consuming. In Quadrant III, decision making and negotiating can be adjusted to the problem or opportunity at hand without the constraints experienced in Quadrant I. Decentralized decision making speeds up the process and negotiating is simpler because the parties involved are more familiar with the organizational characteristics influencing the potential success of adopting and implementing a new technology. Perhaps the overriding factor influencing the acquisition and adaptation processes in the M-type firm will be the technology's potential impact upon the division's profitability.

Communication Systems

A communication system typically evolves out of the authority structure and negotiation strategy used by the company. Without communication, of course, there can be no innovation — this is the philosophy behind "think tanks" and "brainstorming" — so it is not surprising to discover that the quadrants that feature the most restricted or atrophied communication systems are also the least innovative. In Quadrant I, where organizational choice is low and environmental determinism high, communication is mostly vertical and backed up by numerous regulations and conventions that must be observed. Going through proper channels is very important in an organization in this quadrant; low-level employees, for example, are encouraged to communicate with their immediate superiors but are discouraged from communicating with higher levels. The German mail (DP) fits into this quadrant, and the rules governing the vertical communication system are enforced by various powerful unions. In Quadrant II, vertical communication is still the norm, but junior employees are encouraged to communicate and consult with their seniors and management. This freedom is even greater in a Quadrant III organization like Louis Vuitton-Moet Hennessy, in which the communication system, although still governed by certain guidelines, is set up in such a way as to allow for communication across all levels. In a Quadrant IV organization like the Swiss mail (PTT), communication opportunities are seldom restricted but are rarely used. Communication is mostly contained within one's immediate work group, which probably has a lot to do with the fact that firms in this quadrant do not develop their strengths or niches for their products (see Table 3.1).

Summary

As we have seen, there are close links between organizational characteristics, the adoption of innovation, and the accompanying technology-induced organizational change. This chapter's examination of the internal characteristics of the organizational types presented in the previous chapter should help to define the four quadrants. Based on what we have learned, we can safely predict that an organization that tries to move from Quadrants I or IV to Quadrants II or III will find the process impossible unless it either adjusts its internal structures accordingly (Burns & Stalker, 1961; Van de Ven & Ferry, 1980) or reshapes its environment. Because reshaping the environment is beyond the capabilities of most organizations, altering internal structures and processes is, realistically, the only way most can change for the better.

This chapter suggests that the firm must first make a decision about either adopting a new technology or refraining from doing so. After that decision has been made, organizational adaptation must begin to prepare the structure, work processes, and other systems for the imminent change. The success of the adaptation efforts in part controls how soon after installing new technology the firm may benefit. Depending upon the choice and determinism experienced by the firm, the adaptation process may, however, be very different.

An organization must respond to external and internal pressures before a successful adaptation process can be put in motion. Specific levels of environmental determinism and organizational choice experienced by firms delineate certain general and internal characteristics needed to expedite the technology-induced organizational change process (see Tables 4.1 and 4.2). But situations may arise when the incompatabilities between the nature of the technology and the firm's existing configuration of interests, structures, and resources remain despite all organizational efforts to resolve the problem. What is yet to be taken into account?

Although a lot of effort has been expended to increase our understanding of technology management, and especially technology adoption and technology-induced organizational change, previous typologies used to study the phenomenon have not been terribly helpful. One reason may be that changing a firm's quadrant position (see Figure 2.1) necessitates a great many changes in both its internal and its external characteristics, some of which it may be difficult to account for or predict. Organizational factors such as human resource strategy and internal labour market determinism must be considered in any attempt to alter the strategic position of a firm. For instance, moving from one quadrant to the next may require changes as fundamental as adjusting or even rewriting job descriptions and classifications. However, extensive labour market determinism may hinder such broad changes. Furthermore, employees will have to be encouraged to develop new behaviours if the technology is to be integrated effectively into the organization. To reinforce these behaviours, it is likely the firm's reward system will require adjustment, a process that places strenuous demands on strategic human resource management.

Part II explores this subject in greater detail, focusing on human resource management strategies and internal labour market structures—variables that have a great deal of influence on internal structures and that must be taken into account when an organization attempts to realign itself. The relationships between the internal labour market and the authority structure, work organization, and tasks of the firm are discussed, as are the effects of human resource policies upon change and adaptation. Because a fit between organization-environment and internal processes and structures depends largely on internal labour market determinism and the degree of choice in strategic human resource management, a better knowledge of these factors is needed to increase the organization's responsiveness to technology-induced change. This is the subject discussed in Chapter 5.

PART II

Organizational Systems Affecting Technology Acquisition

In Part I, a general framework for studying technology adoption and technology-induced organizational change was described, and we emphasized that environmental and organizational factors must be compatible if the adaptation process is to succeed. Moreover, the interrelationship between environmental determinism and organizational choice was diagrammed to show the role these two dimensions play in successfully fitting new technology to a firm. The internal and external characteristics that tend to accompany varying levels of these two factors were also discussed.

Part II moves on to combine topics that are often separated in contemporary treatment of technology management. It deals with the ways in which organizations may relate to their environments and with the determinants of strategic human resource management, technology adoption, and technology-induced organizational change. The internal labour market, with its rules and regulations, influences the firm's available options and strategies in the human resource domain, thereby affecting the way new technology is introduced into the system. External forces shape internal arrangements and vice versa.

For the year 1987, the German Association of Car Manufacturers (the *Verband der Automobilindustrie*, or VDA) estimated that only 34% of their industry's manufacturing capacity was used. The VDA wished to increase this percentage to 44%, which would have allowed its members to adopt profitable new production technologies for greater efficiency and a substantial reduction in production costs, thus enabling the industry to price its products more competitively. However, the firms could not improve production without increasing the flexibility of work scheduling, and this would have meant rotating shift work, which the industrywide contract between the association and the metal workers union did not permit. Because the association was unable to change the union's stance, the result was that new production technology was not adopted, thereby possibly endangering the industry's competitiveness (e.g., product quality and pricing) in world markets.

This example illustrates the ongoing conflict between *strategic human resource management* and the *internal labour market*. An organization's approach to the former and the influence of the latter will determine the way an organization adapts to new technology. In any given organization, the influence of either variable may be equal to or greater than the influence of the other. The elaborate rules and regulations enforced by the internal labour market (e.g., reclassification of jobs, new job descriptions, and retraining considerations; see also Chapters 8 and 9) can be great enough to slow the decision process needed before adopting a technology and can have a similar effect on the organizational adaptation process required for making effective use of new technology that might actually increase manufacturing capacity (which was the case in our example). A greater level of choice in strategic human resource management, however, could override any of the concerns of a relatively weak internal labour market (e.g., job security and remuneration).

The influence of the internal labour market and of the firm's choice in strategic human resource management, as the example of the car industry should make clear, is largely decided by environmental determinism and level of strategic choice. An industrywide contract is an environmental factor that makes the internal labour markets of organizations in this industry quite strong. This, combined with the lack of choice these organizations experience, limits the kind of human resource strategies that these companies can implement and dictates that such matters as remuneration packages and employee career development be handled in specific ways that the organization will not necessarily consider preferable.

The interrelationship between environmental determinism and levels of strategic choice, and internal labour market determinism and strategic human resource management, makes this latter pair of factors extremely useful in the study of how the former affect technology acquisition and organizational adaptation. With this in mind, Chapter 5 develops four different organizational types based upon the organization's human resource management strategy and the degree of internal labour market determinism it experiences, and then examines how internal systems such as organizational and human resource cultures differ according to these types. The remaining chapters in this part concentrate upon how internal labour market determinism and strategic human resource management affect more specific organizational systems: Chapter 6 focuses on alternative employment strategies and recruitment and selection systems; Chapter 7, on socialization and appraisal systems; Chapter 8, on training and development systems; and Chapter 9, on compensation systems.

5. Internal Labour Market Determinism and Strategic Human Resource Management

As we noted in the last chapter, most organizations trying to improve their relationship with the environment can only do so by first reshaping their own internal structures. The internal labour market (ILM) and strategic human resource management (SHRM) affect both the firm's structure and the firm's ability to change it and influence decision making within the firm. These factors and their interrelationship will in part shape the organization's interaction with its environment. Moreover, SHRM and ILM determine how the firm will respond to environmental constraints and opportunities.

We begin our study by examining the emergence of distinct internal labour markets within organizations and by defining *internal labour market determinism*. We then direct our attention to strategic management, human resource management, and strategic human resource management, and the functions they serve. This will be followed by a contingency model for obtaining a fit between SHRM and ILM that distinguishes four categories of organization. The model indicates what type of organization will experience given degrees of ILM determinism and SHRM and show how ILMs and SHRM can be matched. The four categories of organization are then discussed in terms of the types of internal systems they are likely to feature.

Internal Labour Market Determinism

An internal labour market is defined by the set of administrative rules and procedures that govern product pricing, allocation of human resources, and training decisions of the organization. Its essential feature is its *allocative dimension*, which it uses to allot human resources to positions within the organization based on rules and regulations, which necessitates internal promotion as well as limiting the number of ports of entry into the organization. The allocative dimension could mean that seniority, for example, might affect an individual's chances for promotion more than competency. The

ILM structure affects employees at all levels. The staffing of senior positions in law firms (and also in consulting and accounting firms) through internal recruitment by promotion rather than through external recruitment — a phenomenon documented in research (Wholey, 1985) — exemplifies the way the ILM works for high-level employees.

The factors influencing the formation of ILMs have not been dealt with to any great extent in the literature, but we do know enough about them to name several. *Unionization* is one of the more important (as we have already seen in the example in the introduction to this part); *organization size* is another, larger firms are more likely to have ILMs than smaller ones (Pfeffer & Cohen, 1984). Most researchers also see an association between the *economic sector* and the employee/employer relationship. It has been argued that the nature of the worker's relationship with the organization emerges from the structure of the dual economy (see Osterman, 1984). This claim is supported by Pfeffer and Cohen (1984), who found firms in the core sector of the economy (usually dominated by a small group of corporate Goliaths commanding staggering resources) more likely to have ILMs than firms on the periphery. They also discovered that the structures of ILMs varied according to such variables as whether or not an established promotion-from-within policy existed, whether or not jobs were filled primarily from within, and whether or not prior work experience was required for entry-level jobs that had some promotion potential.

One primary reason for forming an ILM is that managers and workers alike wish to reduce uncertainties arising from the market environment while simultaneously trying to maximize opportunities for profits and wages. From the managers' perspective, one of the most troublesome of these uncertainties is whether or not employees will stay with the firm. This is a concern because although workers come to the firm with certain educational credentials and skills, the organization often invests considerable funds in on-the-job training. To ensure that the firm itself, and not some other organization, receives the returns on its investment, it will take certain steps to reduce the likelihood that its workers will take their skills out into the external labour market. Such steps may take the form of regular incremental wage and salary increases and promotions along internal job ladders, regulated by principles of formal bureaucratic rules (Stark, 1986). These routine procedures are designed not only to bind skilled employees to the firm but also to facilitate the transmission of skills among workers themselves and provide long-term incentives to increase the performance of workers in the production process.

From the employees' perspective, the major uncertainty is that unfavourable external labour market conditions such as high unemployment may drive down the costs of their dismissal, facilitating their replacement by new workers. Thus workers also have an interest in institutionalizing the administrative rules that govern internal job ladders and the wage and salary increments correlated with promotion, and they will attempt to use the influence of their trade unions to establish the aforementioned allocative dimension (Pfeffer & Cohen, 1984). To the extent that they are successful, uncertainty of employment is differentially distributed across the work force and the actual or potential beneficiaries are more likely to cooperate in the production process (Stark, 1986).

One of the unfortunate results of these efforts by both workers and managers to reduce the threat of an uncertain external labour market is the increasing bureaucratization of staffing and rewarding, both of which are internal mechanisms of the firm. This in turn strengthens the ILM, which feeds on bureaucracy (Osterman, 1984, pp. 2-6). We have seen the result of this process in the example of the German auto industry, in which industrial relations are characterized by narrow job classifications organized on well-defined ladders regulated by formal governance procedures.

An organization with a particularly influential ILM may find itself in something like a bureaucratic straitjacket; research has shown that rules can stifle risk-taking and innovativeness by preventing employees from taking chances in their work (Gattiker, 1988a; Gibbs, Keen, & Lucas, 1988). And, as we saw in the last chapter, when individuals cannot be shifted easily from one job to another within the firm, and when job duties and skill requirements cannot be changed without major adjustments of job description and union bargaining contracts, the adoption of new technology is likely to be difficult. It is limiting forces such as these that are referred to when this text speaks of *internal labour market determinism* (ILMD).

Strategic Human Resource Management

An organization's approach to strategic human resource management is based on the decisions made over time by management regarding personnel and activities relating to them. To understand SHRM better, it is useful to define *strategic management* and *human resource management* first.

Strategic Management

The word *strategy* has become extremely overused in recent years, often indiscriminately tacked on to terms in an attempt to add importance or significance to a variety of topics. Derived from the ancient Greek *strategos*, meaning "the art of the general," *strategy* has military roots. In its original military context, the word has been defined as "the employment of the battle as the means to gain the end of the war" (von Clausewitz, 1832/1968). In a corporate setting, strategy could mean the implementation of a management scheme for achieving corporate ends, having to do with determining the basic objectives of an organization and allocating its resources (e.g., time, financial resources and human capital) to their accomplishment.

The purpose of *strategic management* is to create and maintain systems that enable management to react appropriately in any given situation and that facilitate organized action by the firm. As outlined in Chapter 2, the term refers to the decision-making process that focuses the capabilities of the firm upon the opportunities and threats it faces in its environment (Rowe, Mason & Dickel, 1982). To illustrate, recent deficits in the province of Alberta in Canada have led to cutbacks in the budget of the health care ministry. The provision of an undiminished high standard of health care services at a lower cost has, therefore, become the primary objective of this ministry. To accomplish this, the government's strategy involves keeping labour costs down, and this in turn dictates the ministry's stand in contract negotiations with nurses and paramedics.

Human Resource Management

Human resource management (HRM) typically involves the attraction, selection, retention, and utilization of human resources to achieve an effective work force and the attendant individual and organizational objectives (e.g., Cascio & Awad, 1981, p. 3; Werther, Davis, Schwind, Das, & Miner, 1985, chap. 1). It is usually the province of the personnel department. *Labour relations* — the negotiation and administration of union contracts — is a key activity of HRM. HRM is also concerned with performance appraisal; a system may be developed for measuring this on a regular basis for a variety of jobs within the organization. Performance appraisals can indirectly provide the firm with the quantitative data necessary to allow strategic choices and the effective management of unsystematic risk (e.g., labour strikes).

Career management may also be included under the umbrella of HRM. This involves providing career planning and counselling, assisting individuals with self-assessment, making career and job changes possible, and managing both organizational career paths and the internal work force. Career management is a function that tries to make the employee's organizational tenure as productive as possible for both sides, and the other activities of HRM affect it in a variety of ways. The selection of employees, for example, determines the kind of development and career planning that will be necessary to maintain fruitful relationships between management and labour in the future (Gattiker, 1985a). Career management can affect HRM as well, by forecasting possibilities for human resources affected by factors such as technological change and by providing human resource managers with information about future training needs and future demands for employees and skills — information that gives executives an idea of what to look for in external and internal labour markets when it comes time to satisfy these demands in a cost-effective manner.

Strategic Human Resource Management

Strategies for human resource management are based upon the firm's conception of how its objectives can be best achieved with its current human resources in the face of various opportunities and threats. Strategic human resource management should, quite naturally, be seen as beneficial to the company's future growth and prosperity. The activities of SHRM take place over a period of time that can be described as mid- (three to five years) to long-range (five years and up), in contrast to the activities of normal HRM, which are either daily operational functions or short term (up to one year; Cascio & Awad, 1981, chap. 1). Because SHRM is undertaken at one time in order to achieve certain outcomes at a later stage, it is based on assumptions rather than facts — assumptions about the behaviour of employees, the relationship that binds them to the firm, and the company's power to change that relationship (see Mitroff, 1983, p. 38). One of the most important of these assumptions, usually made by organizations that have more effective SHRM, is that employees are important stakeholders in the organization and the organization requires their input and should in turn provide them with satisfactory output (e.g., remuneration). However, the assumptions that guide an organization's approach to SHRM will not necessarily be correct, especially because the information that management uses to make these assumptions is often limited and also because the behaviour of employees and their relationship with the firm may undergo unforeseen changes over a period of time. When these assumptions prove false, conflict between SHRM and ILMD is likely to occur.

Contextual contingencies, such as laws imposed by government upon business, market regulations, and trade restrictions, are *external* factors that, as outlined in Chapters 3 and 4, affect the organization's degree of choice. SHRM and ILMD are *internal* contingencies that affect the management of new technology and especially organizational adaptation. Below, the ILM and HRM contingencies will be discussed in more detail. Contingency theory will be applied to better explain the phenomena of technology adoption and technology-induced organizational change.

Internal Labour Markets, Human Resource Management and Contingency Theory

Structural contingency theory, the dominant theory in the study of organizational design and performance for the past 20 years, is based on a "law of interaction," which states that the performance of an organization is dependent upon a fit between the firm's *structure* and *processes*, given that normal assumptions hold about the premises, boundaries, and system states derived from the theory (see Drazin & Van de Ven, 1985; Dubin, 1976). To achieve effective strategic human resource management, the firm must attain a fit between ILMD and SHRM objectives. *Fit* is an underlying congruence between the firm's SHRM and its ILM. As it is used here, *fit* assumes that an internal consistency of multiple contingencies needs to be achieved to ensure firm performance.

To accomplish a fit between ILM and SHRM contingencies, six structural elements in the human resource domain must be considered: (a) *specialization*: the employees' acquisition of different skills needed to perform work activities attached to a specific position and/or job category; (b) *skill specificity*: the degree of transferability of firm-specific training to other employers; (c) *promotion ladders*: clearly and formally defined pathways that provide promotion opportunities within a firm; (d) *lateral mobility*: interorganizational transferability dependent upon the individual's willingness to perform different tasks and acquire new skills required in his or her work; (e) the *personality assessment* and *selection system* used by the firm with its human resources; and (f) the firm's *reward* and *remuneration system*. The structure defines the makeup of the ILM and suggests approaches to SHRM; ILMD and SHRM in turn provide feedback that redefines the structure. Processes are human resource strategies developed to facilitate this interaction between the structure, HRM, and the ILM.

Given that SHRM and ILMD play such important roles in determining structure and processes, it is easy for us to see how organizational performance will ultimately be contingent upon a fit between the level of ILMD and the amount of choice in SHRM that the firm enjoys. If all relationships between structure and processes and levels of ILMD and SHRM choice were

ILMD and SHRM 81

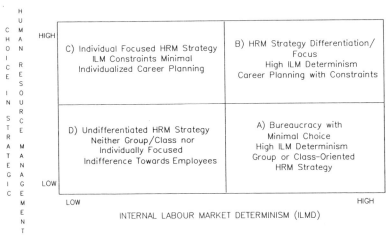

Figure 5.1. Relationship of Choice in Strategic Human Resource Management and Internal Labour Market Determinism

fixed so that a change in any one resulted in the appropriate changes in the rest, processes would always be appropriate to the structure. However, although it is impossible to change structures without changing ILM configurations (and vice versa), a change in structure only *suggests* a change in process or in HRM strategy — it does not *necessitate* it.

Process is defined here as a coordination mechanism used by the firm to facilitate the fit between its HRM system and the ILM. The organization must be self-evaluating, frequently assessing process and deciding whether or not strategic milestones have been achieved and are within the desired parameters of SHRM and ILMD. Because the organization is self-evaluating, it does not always choose the kinds of processes and HRM strategies that the ILM and structures would seem to suggest. If it chooses incorrectly, and there is no fit between structure and process, there is less chance of there being a fit between ILMD and SHRM. This, according to contingency theory, will lead to poor performance.

Figure 5.1 demonstrates the human resource strategies that will most easily allow the organization to obtain a fit between the levels of ILMD and choice in SHRM it experiences. Although the quadrants presented here are not necessarily analogous to the ones presented in Chapter 3 — an organization in Quadrant I (natural selection) of Figure 3.1 would not necessarily be found in Quadrant A (SHRM with minimal choice) of Figure 5.1 — inconsistencies in location between the two figures, if they do exist, are most likely to originate in conflicting contingencies. A firm dealing with hazardous wastes, for example, is likely to experience a high level of environmental determinism in the form of government regulations. Although this necessitates its placement in Quadrants I or II in the first model, the relative newness of the industry might mean that labour would be less organized in this field. There might not, therefore, be a correspondingly high level of ILMD, a situation that would place the organization in either Quadrant C or Quadrant D in the second model. Normally, however, a high level of environmental or structural determinism implies a high level of ILMD, which is to say that the organization affected by extensive government regulation will also be affected by a strong labour presence.

In Figure 5.1, Quadrant A basically shows the conditions or assumptions made for a bureaucratic organization experiencing very limited choice. You will usually find firms in highly regulated markets in this quadrant. These organizations have very little choice in creating human resource strategy because this is mostly controlled by outside influences and by the ILM, which dictates the price for labour (for which the demand is not elastic). Firms under these conditions must either adapt to internal constraints or face the prospect of a strike or some similar crisis. Detailed job classifications and numerous bureaucratic rules influence employees' career paths significantly.

Most railroad companies, for example, are typically found in this quadrant, and using the example of a train engineer is a particularly effective way of demonstrating the restrictions of this quadrant. Because we can imagine that certain kinds of individuals would be more suited to this type of skilled work than others, it is logical to expect that railroads would like to have a certain amount of freedom in their selection of candidates. However, these companies are restricted in their hiring practices by systems in which promotion ladders are clearly defined and intraorganizational mobility is limited. Because of this, recruiting the most suitable candidates is often impossible, and rail companies, barred from recruiting outside the organization by the ILM's allocative dimension, are forced to retrain current employees whose main qualification may be tenure and not talent.

We usually find organizations with both substantial levels of ILMD and high degrees of choice in SHRM in Quadrant B. Examples of these kinds of organizations are Sony or Matsushita, both of which feature powerful ILMs. However, because both of these organizations feature internal company unions, and industrywide contracts are not negotiated, they thus enjoy more freedom in choosing HRM strategies than might otherwise be the case. One method of moving from Quadrant A into Quadrant B is illustrated by Burlington Northern, a U.S. railway that tried to force its unions from industrywide pattern bargaining to direct negotiations with its management. This would have given the organization more choice in SHRM and freed it from having to participate in the sort of industrywide bargaining that, before 1989, resulted in wage paid-miles of service ratios that were extremely high in comparison with similar ratios in the trucking industry, its most important competitor.

Firms in Quadrant C, which is marked by a high degree of choice for SHRM and a low degree of ILMD, enjoy much more encouraging conditions. In this quadrant, there are substantial opportunities for accomplishing a fit between context and structure (see Drazin & Van de Ven, 1985). The firm's autonomy and control in setting SHRM policy are the rule rather than the exception, and the relatively weak ILM has not yet set up exit or entrance barriers. The firm has the opportunity and choice to set up remuneration and reward systems that are geared towards the individual needs of employees, rewarding individual rather than group performance (e.g., Lawler, 1981, chap. 6). Selection and assessment of personnel then focuses on ability, skill, and potential performance rather than on membership and contract obligations. The organization attempts to accommodate individual as well as organizational needs. This quadrant includes firms in many different niches, each characterized by a different set of constraints, opportunities, and competing organizations.

Quadrant D is relatively stable, yet despite a paucity of ILM constraints, firms here tend to lack SHRM choice. HRM strategies are neither group/class focused nor individually focused. These kinds of firm are usually forced to implement SHRM if they seek to increase their competitiveness (Link & Tassey, 1987).

Organizational Types and Organizational Systems

Table 5.1 discusses some of the organizational systems that are characteristic of companies found in the four different quadrants presented here. It should give us a better indication of some of the problems that varying levels of choice in SHRM and ILMD can create.

Organizational and Human Resource Cultures

Organizational culture can be described as consisting of the basic assumptions and beliefs that are shared by all members of the same company (Schein, 1985, chap. 1). These assumptions and beliefs, although they operate at the unconscious level most of the time, define how the organization sees itself

TABLE 5.1: Four Human Resource Strategy and Internal Labour Market Types: Organizational Systems

Variable	Quadrant A *High ILM Determinism, Low Choice in Strategic Human Resource Management*	Quadrant B *High ILM Determinism, High Choice in Strategic Human Resource Management*	Quadrant C *Low ILM Determinism, High Choice in Strategic Human Resource Management*	Quadrant D *Low ILM Determinism, Low Choice in Strategic Human Resource Management*
Organizational culture = primary strategy	Bureaucratic/low morale	Winner/structured	Winner/freewheeling	Cooperativism
Human resource culture = secondary strategy	Socialistic/apathetic	Individualistic/caring	Formalized, performance driven	Traditional
Compensation systems = implementation mode	Average	Good	Excellent	Poor
ILM culture = primary strategy	Class struggle	Cooperativism	Participation	Potential for exploitation
Labour goals = secondary strategy	Maximize number of employed members	Focus on equity and justice	Advanced firm and worker benefits	Indifferent
Job classification systems = implementation mode	Narrow	Average	Broad	Open and linked to hierarchic level

and its environment and may be reinforced by a reward system. The *human resource culture* essentially summarizes the managerial approaches and attitudes towards personnel that explain the state of the organizational culture. Sometimes (especially in successful organizations) the company will have specific strategies for the maintenance of its cultural identity in terms of prevailing values. In a less successful company, on the other hand, the absence of such a strategy will usually go a long way towards explaining why the culture is as undesirable as it is and why it is maintained in such a state.

The most favoured organizational culture is probably that characteristic of Quadrant C organizations, which has values idealizing the collective experience of winning (see Gagliardi, 1986). Consequently, these organizations possess human resource cultures that are strategically geared towards winning as well. These firms are success- and performance-driven, and the limited determinism of their ILMs may lead to the belief that human resources are expendable. Companies developing microcomputer software, such as Ashton-Tate and Lotus, tend to be exemplary of firms in this quadrant.

The work force in Quadrant A, in contrast, possesses one of the least favoured organizational cultures and may have little cultural identity or identification with the firm. The beliefs and behaviour models characteristic of firms in this quadrant invite bureaucracy and losing attitudes (Clarke & Darrough, 1983). This is typically the result of an apathetic human resource culture, which is in turn the result of managerial practices and policies that lack concern for people and that are indifferent to their performance. Such apathy often reflects the general state of demoralization and cynicism that permeates firms with inept or alienated leadership (Sethia & Von Glinow, 1985). Cultures in this quadrant are often socialistic in that they usually feature policies designed to "take care of the employee," policies that are not so much the result of genuine goodwill on the part of the management as they are the outcome of extremely strong ILMs. The earlier mention of Canada Post may serve as an example. Although postal workers might be considered well-off by many people's standards, the company's apparent lack of real concern for its employees in turn causes them to feel antagonistic toward it — a situation that helps explain the rash of postal strikes in recent years.

Although Quadrant B organizations are also bureaucratic, morale is higher and attitudes are more positive, akin to that demonstrated in Quadrant C firms. Firms in this quadrant possess formalized human resource cultures in keeping with their high degree of ILMD and are characterized by a great concern for the individual employee and by strong performance expectations. High performance is encouraged by the employee's belief in the company's concern, just as this concern is justified by the high quality of the employee's work.

Quadrant D cultures, although possessing less strict performance standards, are often even more concerned with the well-being of their members and may be almost familylike in this respect. A fledgling cooperative spirit characterizes this quadrant, which is not as cutthroat as some. However, the organization, like many families, may do little to nurture this spirit and may not seek to exploit it, simply trusting that traditional methods or chance will maintain it. The result may be a seeming indifference on the part of the company to its employees; good relations will be entirely up to the temperament of the managerial staff in general. More traditional company cultures, like those of older, established organizations such as Siemens, ICI, Bank of Montreal, and Ciba-Geigy are often of this type (see Pettigrew, 1985).

Compensation Systems

Compensation systems can be the difference between desirable and undesirable organizational cultures and often determine the attractiveness of the organization as a work environment. *Remuneration* is part of the overall compensation package offered by the firm and provides the salary or wage and fringe benefits as required by law. It is usually the major portion of the compensation the individual will receive from the firm in return for his or her services. The *reward system*, a subsidiary compensation system, provides incentives that direct and motivate employee behaviour to the end of improving individual and collaborative performance (Kerr, 1975). Established criteria often must be met before members of the organization can qualify for these incentives (Sethia & Von Glinow, 1985). Individualized compensation packages, such as cafeteria-style fringe benefits, which permit employees to choose from among many alternatives how their financial compensation will be allocated (e.g., unpaid leave, early retirement, or sabbaticals), might be an attractive part of the reward system, but compensation need not always be financial. Prestige, job security, even a key to the executive washroom — these sorts of rewards are the icing on the cake of the remuneration system.

Although compensation is given by the firm to employees for performing given tasks, the quality of their performance may actually prove to be more dependent upon their *intrinsic motivation*. The relationship between compensation and motivation is uncertain, and theoretical positions have varied regarding the effects of both remuneration and rewards on motivation. Deci (1975) has argued that rewards actually have an undermining effect upon job performance, because they cause employees to think that they engage in behaviours to attain rewards rather than because the behaviours are worthwhile in themselves. Deci assumes that people are intrinsically motivated to perform activities that make them feel self-determining and competent, and that certain kinds of reward or feedback can negatively affect motivation by negatively affecting feelings of competence or self-determination. Arnold (1985), however, found that rewards do not undermine intrinsic interest in tasks unless they are situationally inappropriate, in which case they will most certainly decrease the motivation of certain groups of employees. A pat on the head might be an appropriate reward for a dog that has learned a new trick but not for an employee who has brought the company a multimillion-dollar account. Yet even situationally appropriate rewards are not always great enough in magnitude to increase intrinsic motivation. Varadarajan and Futrell (1984) found that individuals with relatively high levels of training and experience and higher-than-average current earnings tend to have higher expectations with regard to salary increases and rewards, and that these expectations are reflected in higher thresholds of what is perceived as the smallest meaningful merit increase.

It is safe to assume, in the case of a firm possessing an ILM, that differences in the reward structures set up for different groups of employees will be reduced and that (more importantly) the size distribution of rewards will be decreased (Quan, 1984). And because the expected outcomes of high ILMD are greater levels of centralization, formalization, and standardization, combined with reduced flexibility and autonomy, employees tend to perceive an increase in the organization's control over their activities and a correspondingly lower level of self-determination. Sherman and Smith (1984) have found that such external constraints can decrease intrinsic motivation. If the organization, because of lack of choice in SHRM, is perceived as approaching the mechanistic end of the organizational continuum, a similar decrease in self-determination will be perceived, with similar results.

In Quadrant A, remuneration is perceived as being average by the market standard, and the differences between the salaries paid to mid- and lower-level management and those paid to nonmanagerial employees are not great because the former are usually not paid for overtime. Job content rewards are hard to find, as are financial rewards for performance. Nevertheless, compensation packages offer one or two attractive features, such as high job security for unionized members and relatively high wages for lower-level employees. Again, Canada Post is representative of organizations in this quadrant.

Quadrant B organizations offer competitive remuneration packages for both salaried and wage-earning employees. As in the first quadrant, the high level of ILMD assures a certain level of job security for unionized employees based in part on organizational tenure, but the organization retains enough choice in its HRM strategy to offer individualized employment packages as well and may even occasionally go so far as to allow employees to schedule their own working hours. Performance-related bonuses are good, but are limited in that they must be within the boundaries of the negotiated union contract. In Quadrant C, rewards are given on the basis of performance and may constitute a fairly large portion of the individual's compensation package, especially if he or she has a lot of initiative. On the other hand, job security may be tenuous because the influence of the ILM in this quadrant is low, and both voluntary (employee-initiated) and nonvoluntary (firm-initiated) turnover are high because high performance is expected. Those feeling pressured may not find the rewards to be worth the constant strain and may elect to leave before they are fired, and those who do not measure up in the eyes of the firm may leave less willingly. An example of an organization in this quadrant would be Procter & Gamble, which offers superior and frequent performance-based rewards that may be monetary or may take some other form (such as an increased level of authority or responsibility). These performance-based rewards do, however, tend to create a high-pressure culture where the individual must perform very well or else be let go.

Quadrant D organizations have poor remuneration and reward systems. Because performance is not stressed, rewards (usually of the job security variety) are given out in this quadrant only for the completion of contractual obligations or as part of a patronage system, a practice that stifles creativity. The Swiss PTT, discussed in the previous chapter, fits into this quadrant, although considering that the productivity of this company's work force is below average and that the rate of absenteeism is above average, remuneration is probably quite good in relation to the actual amount of work hours put in. If organizations in this quadrant wish to make the environment more friendly to innovation, they would do well to develop reward cultures more like those in Quadrant C.

Compensation systems will be examined in greater detail in Chapter 9, but they were included here because remuneration and reward systems are important tools with which to build an organizational culture. Both can help to reinforce the organization's values and beliefs when it wishes to introduce technological innovation and undergo the organizational adaptation required for effective use of new technology. Compensation and rewards were discussed here as they pertain to managing organizational culture, but in Chapter 9 the focus will be on the various types of reward and compensation systems, thus explaining, for instance, how content-based rewards might differ across quadrants.

ILM Culture

The *internal labour market culture* represents the basic assumptions and beliefs shared by members of the firm concerning the bureaucratic rules and regulations about work, job content/classification, and entry/exit barriers. The ILM culture of an organization in Quadrant A is characterized by an employee-management relationship that can be called a *class struggle* (see also Table 5.1). It is probable that part of the firm's work force is unionized and that these unions try to maximize the welfare of their members, striving to increase the pay and improve the working conditions and job security of the rank and file (Werther, Davis, Schwind, Das, & Miner, 1985, chap. 19).

In Quadrant B, cooperation between the ILM and the firm is the distinguishing feature of the ILM culture. Because competitive pressure is high, the organization and the ILM must work together in the interest of survival and future prosperity. The aforementioned individualized compensation packages and work environments also serve to make relations more friendly. Quadrant C firms extend this cooperation into participation. Although the company is not forced to "treat employees nicely," owing to the low level of ILMD, it realizes that its best interests lie in encouraging these individuals to cooperate and become involved with the company. Company-sponsored sports teams and other recreational activities may be part of this strategy, reinforcing the individual's identification with the firm as well as giving him or her the opportunity to communicate with other members of the organization from different departments or hierarchic levels and increase his or her formal or informal participation in the decision-making process. One organization belonging in this quadrant, Credit Suisse, encourages the use of company-owned sports facilities for exactly these purposes. In Quadrant D, however, organizations with a similar lack of ILMD take a different approach. Although the potential for exploitation is ever present, as in Quadrant C, it is not realized because the limited choice and nature of these organizations cause them to adopt rather laissez-faire attitudes towards the maintenance of the ILM culture.

Labour Goals

Labour goals are characterized by different concerns in each of the four quadrants. In Quadrant A, employees (probably represented by one or several unions) try to maximize the number of employed members, but nonmanagerial employees in Quadrant B focus on maintaining equity and justice for themselves, and try, through their unions, to influence the structure of compensation packages in such a way as to benefit members with tenure. In Quadrant C, the overall advancement of the interests of the firm as well as that of its workers become important, but in Quadrant D, the low level of ILMD and choice for SHRM by the organization may lead to indifferent relationships between the company and its staff. Either the firm may not feel in a position to change its bureaucratic and inflexible strategy when it comes to HRM or the employees may not yet have demanded a more individualized approach to SHRM that considers the special needs of various groups of employees.

Job Classification Systems

The *job classification system* provides the firm and the workers/unions with the chessboard upon which human resource strategies are played out. Narrow job classifications, typical of Quadrant A, do not facilitate technology adoption, subsequent organizational adaptation, or the retraining it necessitates, because they are controlled by rules and regulations that can probably only be changed through the expensive and time-consuming task of renegotiating union contracts. At the opposite end of the scale, in Quadrant C, job classifications are very broad and easy to change, allowing individuals to use different hierarchic ladders and enabling the firm to restructure positions according to the changes that time and innovation will inevitably bring to them as well as according to the individual employee's ability to adapt to these changes.

Although ILMD is high in Quadrant B, the organization still has substantial choice with its SHRM. The firm will try to expand the duties in job descriptions as much as possible to keep the scope of the job relatively wide. Potential conflicts and grievances due to new technology are less likely to occur. In Quadrant D, the job classification system is open and linked to the hierarchic levels within the organization structure. Thus jobs at lower levels may be more strictly delineated than supervisory or management positions. However, when new technology requires new skills or changes a job's content, firms in this quadrant do not always move quickly to reclassify the affected positions, leaving employees unsure of the repercussions the new technology may have on their jobs.

Summary

This chapter should help clarify the relationships between organizational activities, internal labour market determinism and strategic human resource management. The relative degrees of influence of the two variables is ever changing, a reflection of their dynamic natures. Attempting to force a change in this balance is risky, especially given the large number of variables that have been shown to affect it; change is often necessary, however, especially when a company finds itself plagued by an inconveniently high degree of ILMD. An organization might attempt to change this balance either by renegotiating ILM features (e.g., entry barriers or job classification systems) to allow for easier transfers and retraining of redundant personnel (Pfeffer & Cohen, 1984) or by improving its SHRM through new individualized compensation packages (Lawler, 1981).

No matter what strategy an organization trying to offset the effects of high ILMD uses, it is likely to have to deal with unions. As we have seen, unions are usually extremely influential in organizations with particularly strong ILMD. It is interesting to note, therefore, that Quadrant C organizations, which are relatively free from unionization, often voluntarily implement human resource policies that are similar to those unions frequently demand. Fiorito, Lowman, and Nelson (1987), for example, found that participation of workers in quality circles and quality of work life programs reduced the likelihood of unionization. Similar results were found for firms that had grievance systems in which production or quality programs were discussed in small groups and in which communications from management included information about the organization's competitive position. Quadrant B firms, which do have high levels of ILMD, are likely to incorporate similar systems within legally binding contracts between the workers/union and the firm, which may explain why these firms feature "tame" labour markets as opposed to the activist ILMs characteristic of Quadrant A. It would seem that a voluntary attempt on the part of the firm to involve the union or employee representatives in the activities of the organization might be the most effective way of achieving and maintaining a high level of choice in SHRM.

If the company is lucky, its attempts to change the balance between ILMD and level of choice in SHRM will move it towards Quadrant C. Resistance from workers and managers affected by such changes may be substantial, however, and the possibility of failure is ever present. A careless organization might even lose its ability to be self-determining and end up in Quadrant A.

While Figure 3.1 outlines the interrelationship between the environment and the organization and its effect upon strategic choice, Figure 5.1 focuses on the human resource and ILMD interrelationship and its effect upon the technology adoption and technology-induced organizational change process. ILMD and limited choice in SHRM may influence the way the firm can adopt innovations and make the necessary adjustments to use new technology effectively. Chapters 6 to 9 outline how the most important personnel management functions are affected by the various combinations of choice in SHRM and ILMD, as illustrated in Figure 5.1.

Chapter 6 further develops the concepts presented in this chapter, showing how an organization's employment strategies and recruiting and selection systems reflect levels of internal labour market determinism and levels of choice in strategic human resource management.

6. Employment Strategies, Recruitment, and Selection

The organization wishing to facilitate its continuous adaptation to technological innovation has several choices regarding the type of employment contracts it offers prospective employees. It also faces a number of choices regarding its approach to human resource recruitment and selection systems. Even though the interdependence of these various systems makes it impossible for the firm to decide on any one of them without also deciding on the others at the same time, it can — because they are organizational systems and are thus controlled by internal labour market determinism (ILMD) and strategic human resource management (SHRM) — expand its range of choices by altering the relative influence of ILMD as compared with the influence of SHRM policy. It is, therefore, to the organization's benefit to know as much as possible about the ways SHRM and ILMD affect these personnel systems. This chapter introduces this subject. To keep our discussion as simple as possible, we begin with a brief overview of the interrelationships between SHRM, ILMD, and employment strategies and recruitment and selection systems. We end with a discussion of the effects of these strategies and systems upon the acquisition of innovation and technology-induced organizational change processes.

Choice in Employment, Recruitment, and Selection Strategies

Employment strategies and recruitment and selection systems are all vital to the organization attempting to carry out strategic human resource management. The selection system in particular must be consistent with the long-term strategy of the organization and attempt to follow procedures that allow both the organization and its employees to meet their goals (Olian & Rynes, 1984). In some circumstances, however, this may be difficult because all of these strategic systems operate within parameters determined by the internal labour market and, as we have seen, not all firms have the freedom of SHRM that might allow them to manipulate (or, at least, keep a check on) the ILM.

Because of its involvement with the governance structures associated with the development of long-term employment relations (Williamson, 1985), the ILM is concerned with protecting or preserving employee skills threatened by technological innovation. To do this, the ILM might dictate a policy on promotion from within that will affect selection criteria and the organization's recruiting strategy, or it might insist upon other regulations that have similar effects, such as closed shop agreements, demarcation lines, and rules regarding seniority, reentry, custom, and practice. In Germany, for example, work councils usually insist that open positions are advertised internally first and that these positions are filled from the internal work force whenever possible. Management is free to choose from the internal queue, but is restricted in its ability to recruit outside the organization (Windolf, 1986). In a situation where ILMD prevents the firm from arbitrarily dismissing older or less productive workers, the company may not be free to lay off workers made redundant by economic downturns or innovation. That the typical U.S. plant pays for about 8% more blue-collar labour hours than are required to perform regular production work during economic downturns (Fay & Medoff, 1985) is ample evidence of this.

Organizations naturally try to reduce labour costs and the influence of the ILM if it is perceived to add to these costs, often either by relocating in areas with less expensive labour (as many subsidiaries of GM and Ford did when they moved parts of their production from Germany to Great Britain in reaction to German labour costs, which were nearly twice as high in the late 1980s) or by automating or mechanizing certain processes (Hall, 1986, chap. 1). Any of these activities can turn the labour market into a buyer's market in which the firm's preferences are clearly dominant. The organization's ability to choose from the labour market will likely lead to a long-term trend of rising recruitment standards as the organization will usually prefer retraining the members of its work force to wasting capital on the training of new ones and will often see a more careful recruiting process as one means of accomplishing this.

Not all firms, of course, will be able to use these methods, especially not those in quadrants with low levels of choice in SHRM. It is, therefore, safe to assume that employment, recruitment, and selection strategies vary according to the different quadrants presented in Figure 5.1 in the previous chapter, an assumption we will test in the following sections.

Alternative Employment Strategies

It is not unusual for an individual to leave an organization after it has invested a great deal of time in his or her training and before it has benefited from that training. This is an inherent risk in recruiting, and many firms, in addition to (or, in some cases, rather than) trying new recruitment strategies they hope will solve this problem, will seek alternative methods of employment as a sort of insurance policy against the failure of these strategies. These alternative methods do not necessarily rid the organization of the expense of training, but they may eliminate other employment expenses or encourage individuals to stay with the company. For this reason, we examine these alternative strategies before we look at recruitment and selection.

Guaranteed Part-Time Employment

Part-time employment is becoming more and more common. Part-time employees in the U.S. work force increased from 13% in 1967 to over 18% in 1984 (Bureau of Labor Statistics, 1982, 1984; *Monthly Labor Review*, 1968, Vol. 91, p. 92). In 1982, Canadian figures also indicated a larger group of women working part-time—25% versus only 6.9% of men (Boulet & Lavallée, 1984). Some countries in Europe report that one out of five individuals works part-time, and up to six times as many women work part-time as men (OECD, 1982). In Switzerland, for example, at the end of 1988, 5.8% of all working men and 34.4% of all working women were employed part-time (NZZ, April 8, 1989). As a percentage of the total working population, part-time employment increased between 1982 and 1988 from 12.5% to 16.1%. Part-time employees are "safer" for the firm than full-time employees in that management has more control over them: Poorly protected by laws or union contracts, part-timers are easier to dismiss when the economy slumps or when an innovation makes their positions redundant. The previously mentioned U.S. plant that pays for more blue-collar labour hours than it requires could, for example, eliminate this problem by employing one out of five workers as part-timers during peak periods and then laying these workers off during recessions. This method of employment, however, has drawbacks that may offset its advantages to the organization. Some research comparing part-time and full-time employees' reactions to work has indicated that part-time workers may process organizational experience differently, often less positively, than their full-time peers (e.g., Miller & Terborg, 1979). In apparent agreement with this notion are numerous studies finding levels of job satisfaction to be significantly related to five different job descriptive index (JDI) facets for full-time workers but not for part-time workers (Jackofsky & Peters, 1987). Voluntary turnover is also affected by part-time employment as the voluntary turnover rates for part-time workers seemingly cannot be predicted from variables commonly hypothesized to precede and predict turnover for full-time employees (Peters, Jackofsky, & Salter, 1981). Apparently, however, this depends more upon the *treatment* of part-time employees in the organization than the actual number of hours they work. A later study by two of the same researchers (Jackofsky & Peters, 1987) found that, when part-time employees are treated similarly to full-time workers (which was not the case in the original study, as it examined firms with policies and practices that distinguished between part-time and full-time employees), similar predictors for turnover are present. We can conclude, therefore, that although part-time employment may provide answers for the organization seeking flexibility, it will not necessarily provide answers for the organization seeking dedicated and enthusiastic personnel. In addition, there is always the risk that the ILM structure and union contracts may define a part-time job as permanent after a certain time. If this happens, and part-time workers must be treated like full-time employees, the flexibility advantage will have been eliminated.

Nonguaranteed Part-Time Employment

Part-time employees who work a guaranteed number of hours a week, although not being as well-off as their full-time coworkers, still usually receive some benefits by law. This is not as likely to be true of part-time employees whose hours are not guaranteed. Because additional wage costs per hour have been reported to range from between 30% to 95% of the hourly wage paid, the company can realize significant savings by hiring employees whose wages are lower than those paid to full-time employees for overtime and who are also not entitled to those costly benefits. A U.S. study showed a link between high, fixed wage costs and the increase of part-time employment (Montgomery, 1988). More than 5,000 organizations were surveyed to examine the ratio of part-time weekly hours to full-time weekly hours at firms employing both types of workers. The results indicated that relative higher quasi-fixed labour costs increase the ratio of part-time to full-time hours.

A study commissioned by the Dutch government (NZZ, January 9, 1988) indicated that several hundred thousand employees in Holland do not have guaranteed, contracted working hours. Many of these people work at home, but a large number—estimated to be about 300,000—are "on call." Most work flexible hours out of necessity and are employed in the hospitality, tourism, retail, janitorial, health, food, and printing industries; three quarters of this group are women. Using this type of employment helps the firm reduce the additional wage costs per hour from approximately 80% of the hourly wage for full-time employees (UBS, 1985) to possibly less than 10% for nonguaranteed part-time employees. Some of the firms surveyed employed as much as 10% of their work force in this capacity, a figure that is, at present, higher than that typical of North America and the rest of Europe.

Although an employment strategy such as this offers the organization freedom from paying unemployment or health insurance and so forth, like guaranteed part-time employment, it also has drawbacks, especially in human terms. The greatest disadvantage of this employment scheme is that it creates two distinct classes of employee: the "privileged," who enjoy job security, fringe benefits, and a regular, relatively stable income, and the "nonprivileged," who do not have regular and stable incomes and whose fringe benefit packages contain the minimum benefits required by law. Although it is true that a number of individuals prefer this sort of "hand-to-mouth" existence — especially those with other part-time careers like actors or artists — the Dutch statistics show that we can by no means assume that these individuals represent the majority. People with families, in particular, who have a greater need for the benefits provided by a guaranteed job, suffer greatly as members of the nonprivileged class.

It is likely, however, that a great increase in the numbers of employees in the nonprivileged class would be accompanied by an outcry against the lack of benefits and an appropriate regulatory reaction on the part of the government. The high number of nonprivileged employees in Holland has already led to government proposals to do away with the systems that allow many kinds of nonguaranteed labour, so the two-class labour market might never come about.

Work Sharing

The basic principle of *work sharing* is that two or more people share one job. Although this neither reduces the benefits the organization has to pay nor limits the wages of its employees, it saves the organization training expenses by giving it an alternative that can be used to encourage individuals to stay with the firm. Certain groups are more likely to favour work sharing than others (Blyton, 1987). Professionals with higher incomes and skills that are in great demand, for example, may wish to work fewer hours and devote more of their time to other interests, and may demand work sharing as a means of accomplishing this, but labourers might tend to be less willing because they already have low incomes. It is perhaps not surprising, therefore, that work sharing is rarely found in labour-intensive industries but is often found in professional occupations in which qualified personnel are scarce. This is true despite the fact that labour-intensive work, which does not often require specialized knowledge or talents, is ideal for work sharing, while the kind of work that is done at higher levels might confound the efforts of individuals trying to share duties. Work-sharing initiatives have not been very successful in most countries, apparently because many employers and governments are biased against them and believe that work sharing will *increase* rather than reduce labour costs, expecting that unions or the ILMD will demand that both job holders receive full fringe benefits. This problem could be solved by having employees share benefits as well as jobs, so that the sum of the benefits paid to both would be equal to the benefits paid to a full-time employee holding a similar position, but a suggestion in this vein would be likely to meet with a cool reception from organized labour. Work sharing under term contracts would eliminate the problem altogether because the full fringe benefits of a permanent position would not have to be paid, but this phenomenon is nearly nonexistent. Even if the organization were to resolve these problems, however, it is difficult to see how work sharing could be of benefit at the common labourer level as these workers can easily be replaced.

Term Positions

Because of the potential flexibility offered by term positions, companies have started to offer more of these while holding the number of permanent positions constant. The financial advantages are similar to those provided by employing part-time labour. Often *term contracts* require employers to pay only legally required benefits like holiday pay and unemployment insurance without having to put money into other programs like dental and pension plans. Term contracts also provide one of the simplest and most palatable means of dismissing redundant employees who could be discharged simply by not renewing their contracts. These contracts can be offered to both full-time and part-time employees and are especially attractive to firms in highly competitive markets that face levels of environmental or ILM determinism that do not allow for the dismissal of redundant workers with seniority.

The educational system in Canada graphically illustrates the benefits of term positions. Canadian school districts and universities are offering non-tenure-track positions with growing frequency. By hiring term position employees only for the period when school is actually in session, they can custom-tailor their curricula and save money during the summer when most students (especially those at the lower levels) enjoy several months off and when permanent staff still receive pay. Because these employees do not acquire the seniority rights that would increase their salaries, they can be kept at a fixed wage indefinitely. This approach represents an appealing alternative to the risky prospect of hiring permanent teachers who are nearly impossible to dismiss once they have acquired tenure (see Piore, 1986).

SEAT (Volkswagen's Spanish subsidiary) is using term contracts (12.5% of its total work force have three-year term contracts) to cope with its worker's union demand to hire new employees instead of asking permanently employed workers to do overtime. Again, once workers are employed full-time, it is nearly impossible to lay them off or fire them (Knipper, 1989), and term contracts are used in this case to attain some flexibility in the firm's work force to cope with business cycles.

Overtime and Flexible Work Scheduling

Perhaps one of the simplest and most convenient alternatives to paying for the training of new staff is to eliminate the need for this staff by encouraging present employees to work overtime. The obvious drawback to this approach is that overtime can be extremely costly as employees will usually be paid anywhere from 1.5 to 3 times their normal wages for each extra hour. Nonetheless, a number of German manufacturing firms found this option more affordable than hiring new staff after a new industrywide labour contract in 1984 lowered the weekly hours for most of its workers from 40 to 38.5 or 37. Other manufacturers, like those in the German Metal Manufacturing Association (IG-Metall), when faced with the same situation, used a different approach and decreased the average number of hours worked by all labourers to the required 38.5 by having certain groups that possessed greater skills work 40 hours and others that were relatively unskilled work only 35. These organizations could thus make more efficient use of the man-hours that were available to them. Many of these firms also implemented flexible work scheduling, contracting their employees to work a certain number of hours per week without stipulating the number of hours they would work on any given day. Employees would then be required to work overtime on days when the demand for their skills was great without receiving overtime rates.

Organizational Type and Employment Alternatives

Organizations usually favour a combination of the alternative methods of employment presented above and retain labour market flexibility by increasing part-time work forces under term contracts or by letting their current staff work overtime more often. This flexibility allows firms to cope with more stringent ILM rules and governance structures and to adopt technological innovation more smoothly. Table 6.1 shows how the four organizational types manage their employment systems.

Part-Time Employment

As the findings of the previously mentioned Dutch study show, companies who structure and plan their production poorly will tend to have more on-call part-time employees than more efficient firms. However, high ILM determinism can force an organization to offer its part-time workers a guaranteed number of hours either per week or per month. An organization in Quadrant A, therefore, avoids hiring on-call personnel. One in Quadrant D is likely often to need such people in addition to those it employs as guaranteed part-timers and has the freedom to make use of them without interference from the ILM. Quadrant C firms, who structure and plan their production well, prefer to hire part-timers with contracted hours. Organizations in Quadrant B, on the other hand, might find it necessary, despite high ILMD, to hire on-call personnel as well as guaranteed part-timers, not because they plan production poorly but to manage abnormally high work loads cheaply. Airlines like World Airways, for example, hire students with foreign language skills for the high-traffic summer season as flight attendants without guaranteed working hours. The flexibility these students provide is seen as desirable because the demand for the service this industry supplies may rapidly diminish as the result of market conditions impossible to foresee, such as hijackings or terrorist attacks that might lead to a temporary fear of air travel.

Term Positions

Although Quadrant A and B firms may wish to use part-time guaranteed (e.g., 20 hours a week, working 4 hours each morning for a specified time) and on-call term employment contracts (e.g., working as many hours as called upon during a week for a specified time but without a guaranteed number of hours) to reduce ILMD, the unions of a Quadrant A firm like

TABLE 6.1: Four Human Resource Strategy and Internal Labour Market Types: Employment Strategies

Variable	Quadrant A High ILM Determinism, Low Choice in Strategic Human Resource Management	Quadrant B High ILM Determinism, High Choice in Strategic Human Resource Management	Quadrant C Low ILM Determinism, High Choice in Strategic Human Resource Management	Quadrant D Low ILM Determinism, Low Choice in Strategic Human Resource Management
Part time				
on call	Rarely used	Used to manage high work loads	Rarely used	Used as needed
guaranteed hours	Often used in part due to ILMD	Used to manage high work loads	Used for employing the services of highly skilled workers	Used as needed
Term employment				
on call	Used to counter negative ILMD effects	Used to counter negative ILMD effects	Used to hire people with know-how for special projects	Rarely used
guaranteed hours	Used to counter negative ILMD effects	Used to counter negative ILMD effects	Used to hire people with know-how for special projects	Rarely used
Work sharing	Not an option	Used if permitted by ILM	Used whenever needed	No
Overtime	Attractive with high ILMD	Attractive with high ILMD	Used as needed	Used as needed
Flexibility in work scheduling	Not really an option unless union agrees or part of the ILM agreement	Option, used only if required by production process	Used whenever advantageous for production and competitive positioning of the firm	Rarely used

Canada Post may not allow the hiring of people under any sort of term contracts without paying them fringe benefits and letting them acquire seniority rights. Firms in Quadrant B may use term employment to hire individuals with special skills or knowledge for the duration of special projects or to improve their competitive position. Firms in Quadrant C may also use term employment for short-term projects; Credit Suisse, for instance, hired specialists for its extensive information system implementation program under special term contracts. Companies in Quadrant D, by contrast, lack the necessary HRM strategy to make effective use of the possibilities offered by term employment.

Work Sharing

As mentioned earlier, work sharing, when used at all, is most often employed to retain highly paid professionals who prefer and can afford to work half-time. Again, firms experiencing choice in their SHRM may use this option to employ individuals who offer unique skills and who may demand such flexibility in work arrangements before joining the firm. This is the case in several European countries, where husband and wife teams sharing one full-time teaching job or one full-time nursing position are quite common. Quadrant A organizations may lack the choice to use these kinds of arrangements; Quadrant D employers might lack the vision.

For Quadrant B, ILMD rules are likely to require the firm to pay full fringe benefits for any permanent position from half-time up to full-time. Work sharing could mean that two people split working hours evenly while both enjoy full benefits. In such a case, the organization may not allow, or at least not encourage, job sharing. Some ILMs, however, do allow the firm to split the benefits between two job holders. In Quadrant C, the ILMD is a minor factor, thus the firm may choose to use job sharing. An example of a case in which it might be utilized is for a female employee who decides after having a child that working half-time may be more suitable to her new life-style. Here, the firm may want to accommodate the needs and wishes of the productive long-term employee by offering job sharing.

Overtime

As Table 6.1 indicates, firms in all quadrants use overtime whenever necessary, although probably for different reasons. Union agreements with firms in Quadrants A and B may make the paying of regular workers for overtime a much more attractive prospect than calling in part-time workers.

A Quadrant D organization with weak SHRM and a deadline to meet may be forced to require overtime. In Quadrant C, low ILMD may allow the firm to compensate employees for overtime with time off rather than extra pay.

Flexible Work Scheduling

Flexibility in work scheduling may be an option for firms in Quadrants B and C because both enjoy a certain level of choice in SHRM, especially those in Quadrant C with low ILMD. As we have already seen, Quadrant C organizations often compensate employees for overtime with time off, which is much the same as flexible work scheduling. In Quadrant B firms, as the earlier example of the industrywide contract with the German metal workers' union has shown, high ILMD does not necessarily prevent a firm from being flexible in scheduling its work force and employing personnel according to the company's need for them. In a Quadrant A firm, such as Canada Post, where high ILMD is combined with limited choice in SHRM, flexibility is also curtailed. A Quadrant D employer such as the Swiss PTT, having little choice in either dimension, is not likely even to consider flexible work scheduling and, if it did, its inability to implement HRM strategies would be likely to cripple its ability to take such an approach.

Although alternative employment strategies such as those discussed above are fairly common, the greater part of most work forces — usually more than 80% — remains composed of full-time employees. Because contract regulations and ILMD may make it difficult for a firm to dismiss full-timers after their probationary periods are over, and because training can place great demands upon the organization's resources, successful recruiting and selection system techniques are essential.

Recruitment and Selection Systems

Recruitment and *selection* are so closely related in most staffing programs that researchers have often avoided drawing distinctions between the two. Nevertheless, for our purposes a distinction seems helpful. In this context, *recruitment* should be understood as defining the activities or practices that determine the characteristics of applicants to whom selection procedures are ultimately applied. Recruitment occurs prior to selection and determines the type of applications from which predictor information is gathered for use in the final hiring process. *Selection*, on the other hand, involves evaluating predictor information for the purpose of making a final hiring decision.

Recruitment and selection systems must be designed to attract individuals who will fit easily into the desired organizational culture (see Table 5.1). These systems can play a crucial role when trying to attract employees with scarce skills in the marketplace or work attitudes appropriate for facilitating technology-induced organizational change. Furthermore, fundamental change in the organization, such as an attempt to move from, for example, Quadrant A (high ILMD and low choice in SHRM) to Quadrant B (high ILMD and high choice in SHRM; see Figure 5.1) requires adjustment in the recruiting and selection systems, which are responsible for hiring the employees needed to promote that change. Potential modifications are further outlined below.

Many activities conducted prior to extending a job offer might appropriately be viewed as either recruitment or selection procedures, depending on how they are used in a particular situation. For example, resumé information can be used either as a recruitment or a selection device. If such information is applied to an applicant pool (either alone or in combination with other predictor information) in order to choose among applicants for final hiring, then it is properly classified as a selection device. If, however, the resumé information serves to identify those who will be evaluated subsequently via other predictors, then it changes the nature of applications evaluated at later stages. In this context, resumé information serves a recruitment purpose.

Recruitment

Because ILMs are established to develop long-term employment relationships and to bind workers to the organization so that turnover and training costs can be reduced, the firm improving its recruiting strategy must do so within the rules and guidelines of the ILM framework. The recruitment process usually begins with a definition of the *ideal candidate*, arrived at through the use of the selection criteria discussed earlier. The ideal candidate is a theoretical individual who would best fit organizational requirements. The recruiting channels a company may choose from are diverse; an individual interested in joining an organization can reply to advertisements in newspapers, in journals, or at conventions, enter college placement programs, or even apply on his or her own initiative. However, one method of recruitment is not always as good as another. Breaugh (1981), in a study using professionals, found that individuals hired through different channels differed in their absenteeism rates, dependability, job involvement, and work quality. Respondents to convention or journal advertisements and self-initiated applicants were found to be the best sources of recruitment for companies. Although these findings seem to discourage companies from advertising positions through certain media, it is worth noting that such differences are not *always* apparent. Swaroff, Barclay, and Bass (1985) found none in a study surveying skilled employees. Taking the findings of both studies together, there is an indication that any recruitment process is as good as any other for soliciting applicants for some — perhaps most — kinds of job but others may require more specialized recruiting. This implies that not all recruitment processes are equally specific. Newspapers, for example, may be consulted by seekers of sales jobs but not university appointments. We do not have to agree with this theory to realize that both studies clearly indicate that a specific job description for an open position and a listing of specific individual qualifications will usually attract qualified applicants, and a vague job description and vague individual qualification requirements will encourage unqualified candidates to apply (Mason & Belt, 1986).

Past work experience and education information, along with other personal history information gleaned from resumés and job application forms, is often used for preselection during the early recruitment process. There are several evaluation methods, such as the point method, the grouping method, the task-based method, and the behavioural consistency method (see Ash & Levine, 1985, for a discussion of these) available to organizations engaged in recruiting. In the public sector (e.g., government agencies), the procedure has been formalized to the extent that judgments are translated into scores used for ranking applicants. Once the person has been hired, similar appraisal instruments will be used to assess performance throughout his or her career (a topic discussed further in the next chapter).

Selection

The organization has a variety of options regarding its choice of selection criteria. It must, therefore, choose carefully, because these criteria shape how all subsequent decisions (e.g., method of recruitment, choice of selection devices) should be carried out for maximum staffing effectiveness. A basic premise of organizational staffing is that selection criteria should be closely aligned with job requirements. In practice, this is usually accomplished through *job analysis* — a procedure that captures the behaviours and activities performed by the present job holder. Recent legal guidelines and court decisions have further reinforced the need for directly linking selection criteria to the requirements revealed by job analysis (see Rohmert & Landau, 1983). For instance, based on job analysis, a sample job may be used to test applicants (e.g., a typing test for a secretary) and job knowledge may be examined (e.g., giving an accountant the necessary figures to produce a consolidated balance sheet within a given time period).

One of the biggest challenges is to make tests valid and reliable. *Validity* means that the test statistically predicts what it is supposed to predict. Thus a person who passes a typing test should be able to perform well as a typist if the test is valid. If later job performance confirms the earlier test results (that is if high test results equal high performance), the test shows predictive validity. The relationship between predictors and performance information indicates the predictor's validity, but *reliability* indicates whether a test measures the same thing consistently.

Selection often involves *personality testing*, although the measurement of personality traits is an inexact science at best. Although traits such as honesty, responsibility, independence, and sociability are usually seen as desirable and even necessary for many occupations, traditional psychological testing devices typically fail to predict associated job behaviour with anything approaching satisfactory rigour (Ben-Shakhar, Bar-Hillel, Bilu, Ben-Abba, & Flug, 1986). The increasing demand for better personnel selection, combined with the weakness of standard personality tests, has led many firms to turn to alternative predictor methods — most notably, graphology.

Graphology involves the creation of a freestyle overall qualitative personality description based on a handwriting sample submitted by the job applicant. Rafaeli and Klimoski (1983) estimated that 3,000 American firms use graphology, and the number appears to be growing. In Europe, graphology is widespread and becoming more so, and Ben-Shakhar and colleagues (1986) report that, in Israel, graphology is more common than any other single personality test. The popularity of this method appears to be unwarranted, however. Validation of the criteria used by graphologists is hard to come by because such traits can seldom be independently ascertained with sufficient certitude, and, according to research, graphologists usually do not perform significantly better than a chance model. Rafaeli and Klimoski, using an American sample, detected no evidence for the validity of graphology.

Although personality testing is often used for professionals and skilled employees, it is not as commonly used for industrial jobs and semi- or low-skilled positions. For these kinds of job, as Reilly, Zedeck, and Tenopyr (1979) have pointed out, tests of physical ability are more important, in both human and economic terms. In human terms, workers who are physically unsuited to their jobs run an increased risk of causing physical injury to themselves and other workers; in economic terms, injuries result in increased medical payments, workers' compensation benefits, and the costs associated with replacing the injured worker. To these expenses can be added costs resulting from the general decrease in productivity of individual employees and from non-injury-producing accidents caused by inept workers (Arnold, Rauschenberger, Soubel, & Guion, 1982).

Tests are not the only predictors of the individual's employment potential. College experience, for example, has often been seen as an indicator of management capability, and the relationships between the two, as demonstrated in assessment centre performance and later promotions, have been studied using longitudinal data sets (Howard, 1986). In Howard's study, humanities and social science majors had the best overall performance when going through a three-day assessment centre and as managers later on, and MBAs seemed to have good administrative and cognitive abilities. Grades, however, were found to be weak predictors for the performance of participants in the three-day assessment centre and for their on-the-job performance. Participants were reassessed after 8 and 20 years, with the overall indication being that, although college experiences can be a meaningful determinant of managerial performance, one needs to relate particular university experiences to the specific criteria emphasized in different jobs.

Organization Type and Recruitment and Selection Systems

We can learn more about the ways ILMD and SHRM influence organizational recruitment and selection systems by examining these systems in terms of the four quadrants presented in Figure 5.1 (see also Table 5.1). The list presented in Table 6.2 is, of course, not complete and should only be seen as a summary.

Recruitment

Quadrants A and B in Table 6.2 illustrate the systems employed by organizations faced with high levels of ILMD. The distinct variance between the two quadrants is the result of different levels of choice in SHRM. Firms in Quadrant A may employ a status quo strategy (i.e., retaining an approach for numerous years because "things have always been done this way," even though it may not be successful in attracting the type of applicants wanted), recruiting individuals who have been recommended or referred by other employees and by community leaders in particular. Firms in Quadrant B show some limited innovative efforts; holding a company night with free beer and pizza at a local university to start off their recruiting campaign is an example of a potential strategy.

TABLE 6.2: Four Human Resource Strategy and Internal Labour Market Types: Recruitment and Selection Systems

Variable	Quadrant A High ILM Determinism, Low Choice in Strategic Human Resource Management	Quadrant B High ILM Determinism, High Choice in Strategic Human Resource Management	Quadrant C Low ILM Determinism, High Choice in Strategic Human Resource Management	Quadrant D Low ILM Determinism, Low Choice in Strategic Human Resource Management
Recruitment	Walk-ins, write-ins, advertising, labour unions, internally, employee and community referrals, educational institutions	Walk-ins, write-ins, advertising, employee referrals, intracompany transfers, labour unions, and educational institutions	Walk-ins, write-ins, advertising, employee referrals, intracompany transfers, educational institutions, employment agencies, professional organizations, and other opportunities	Walk-ins, write-ins, advertising, employment agencies, and other opportunities
	Status quo strategy	Limited innovative efforts	Innovative efforts	Intermittent efforts
Selection	Functionally oriented: subjectively used; personality testing rarely used for employed members	Functionally and generally oriented: systematic criteria used, specific job descriptions and qualifications; personality testing used for supervisory personnel and management	Functionally and generally oriented: systematic criteria used, specific job descriptions and qualifications; personality testing used extensively for predicting training success and job proficiency across all functions and levels	Functionally oriented: limited systematic approach, "muddling through"; personality testing rarely used at any level

An example of a Quadrant A organization is Hollandsche Beton Groep (the largest construction company in the Netherlands), which uses status quo strategies to find new employees and apprentices and must hire from within if qualified applicants are available. Sony is an example for Quadrant B; it favours recruiting recent graduates of the most prestigious universities and intracompany transfers (i.e., being sent abroad to work in a different division, then being called back after several years).

In Quadrants C and D, low levels of ILMD allow firms to recruit through a variety of channels. However, although Quadrant C organizations take advantage of this opportunity and are often innovative in their recruitment processes — for instance, offering summer internships to students that can lead to permanent employment — Quadrant D firms show only intermittent efforts in their recruiting. In all quadrants, interest groups may make their presence felt in the recruiting process, as they may pressure the government to encourage recruiting of minority groups such as women, natives, or blacks.

To illustrate the situations faced by companies in Quadrants C and D, we can look at two Swiss examples. Credit Suisse uses a variety of innovative efforts to attract employees, even going as far as offering prospective employees part-time and summer jobs while they are still attending university or high school. The company may also recruit directly through newspaper advertisements, employment agencies, and "headhunters" (a professional placement coordinator who tries to find executives for job openings and jobs for executives). In contrast, the efforts by the Swiss PTT are intermittent and do not always give an outsider the impression of a clear recruiting strategy. Both walk-in and write-in applications are treated in a standard fashion(e.g., "complete an application and we'll be in touch").

Selection

In Quadrant A, an organization may have a subjectively used selection process. Just as referrals and seniority rights play a large part in the recruiting of individuals, they are important in the actual selection of employees, far more so than personality testing. In Canada Post, for example, seniority determines not only who is eligible for front-counter positions in post offices but also who actually gets these positions, even though the customer contact that is an important part of these positions seems to require interpersonal skills that personality testing might reveal. Qualifications other than seniority are only considered if several internal applications have been received from employees with the same tenure. It is possible that one of the reasons personality testing is rare in this quadrant is that employers here have neither developed the skills to interpret the results of personality tests nor agreed upon their usefulness in helping select and assess individuals. Because the validity of such tests is limited when trying to predict, for instance, an employee's training success and job proficiency, unions in particular may resist the use of such techniques (e.g., Hogan, Carpenter, Briggs, & Hansson, 1985). Functional orientation in this context means that selection is solely directed towards finding a fit between job requirements and the individual without necessarily also considering future career development and growth needs of that individual.

Firms in Quadrant B are different in that they are able to use various systematic criteria and clear job descriptions to choose employees, although the high level of ILMD and seniority rights similar to those in Quadrant A may deter the organization from using personality testing on nonmanagerial employees. The selection is generally oriented, which means that they consider not only the job requirements but also future growth potential and fit between the organizational culture and the individual. Selection criteria and tests must be particularly reliable, valid, and effective if the firm does not wish to incur the wrath of the ILM. An organization in this quadrant, like Canadian Airlines International, might not, therefore, use personality tests for its maintenance workers, although it would use them for supervisory and management personnel.

In Quadrant C, organizations may use systematic criteria and job descriptions for potential job holders and do not have to give internal applicants preference over external ones. Using methods such as assessment centres, the firm can make extensive use of personality tests, among others, in the hope that they will provide an accurate personality profile of the potential job holder. Quadrant D firms, on the other hand, do not have the foresight and human resource strategies to carry out testing. Their approach to selection is as muddled as their approach to recruitment—barely systematic and functionally oriented. Thus selection may appear somewhat arbitrary to the outsider and be mostly based on current requirements without consideration of the future needs of the organization or employee.

Facilitating Technology-Induced Organizational Change

As mentioned earlier, adopting new technology and undergoing the necessary organizational changes so that it fits the firm requires consistency not only between external factors (e.g., government regulations and market demands) but also internal systems such as personnel recruitment and selection.

To illustrate the links, consider a company that has a substantial group of employees on term contracts. To absorb productivity increases achieved with the help of new technology, such a firm can reduce its work force, without angering its internal labour market, simply by not renewing some of these employment contracts when their terms expire. For an organization enjoying high ILMD (Quadrants A and B), however, which is unlikely to have many employees on term contracts, this is not an option. To attain a reduction in the work force, companies in these quadrants may have to offer substantial severance packages to their current full-time employees. Additionally, the choice of which employees to lay off may have to be based on tenure rather than on such factors as performance and career potential.

Radical innovations, and their rapid acquisition by the firm, might make it difficult to train the work force adequately. The firm may, therefore, try to acquire the necessary skills by recruiting people from the outside who have the required skills. High ILMD may limit this option somewhat, forcing a firm to list the jobs internally first and then provide interested selected employees with the necessary training. If this is the case, these inevitably time-consuming procedures may defer the introduction of the technology and, as a result, reduce the competitive edge the company hopes to gain by being an early adopter.

Strategic human resource management can be instrumental in changing organizational strategy from reactive to proactive. Recruiting efforts must reach the most desirable target population, and selection procedures must enable the firm to hire the most suitable individual for the job (although it is becoming increasingly difficult to attract and retain high-calibre personnel). As it is, the employees who ultimately determine the fate of any efforts to adopt innovation (Bernardin & Beatty, 1984), people who really fit the corporate and human resource culture, must be recruited, selected, and retained to facilitate technology-induced organizational change (Kimberly, 1981).

Technological innovation and further internationalization of trade is destined to increase the demand for better utilization of human resources. Employment options such as part-time employment are likely to increase — even now, the demand for this type of work exceeds supply by a factor of about two (e.g., Switzerland, NZZ, April 8, 1989). Moreover, further reductions in the average workweek will increase the demand for higher productivity. For instance, although it was IG-Metall's intent to make more jobs available when obtaining a reduction in the workweek from 39 to 37 hours for its members in 1984, this increase in employment did not occur. Instead, management introduced more technology in an effort to raise the productivity of those workers already employed. IG-Metall then announced in 1989 that it would strike in 1990 to obtain a further reduction to 35 hours per week and to procure better fringe benefits and wages for members (NZZ, December 14, 1989). Such developments further encourage firms to look for new employment options to reduce additional wage costs and allow for greater scheduling flexibility.

The more unions and employees bargain through the ILM and industry-wide contracts for increased benefits, job security, and wages, the more likely it is firms will try to avoid the full impact of these developments. Unfortunately, a balance between these conflicting tendencies is nowhere in sight.

Summary

From what we have learned in this chapter, it is easy to see how the high levels of centralization, formalization, and standardization, and the reduced levels of flexibility and management autonomy that high internal labour market determinism implies, can make the existence of a human resource management strategy focused on individuals a near impossibility and also make it very difficult for the organization to select individuals according to their intrinsic worth. Environmental determinism in the form of industrywide agreements is especially influential in determining the sway the ILM may have in these matters. However, the potential to overcome these handicaps does exist. The firm may create employment strategies (as Table 6.1 demonstrates) or develop tools useful to the recruitment and selection of qualified job applicants (see Table 6.2), both of which can reduce the control of the ILM in these matters.

Environmental factors such as competition necessitate acquisition of new technology by the firm to remain viable. Nonetheless, technology-induced organizational change may be greatly aided if employment, recruitment, and selection strategies allow for change. ILMD may restrict such efforts, however, hindering technology-induced organizational change and ultimately leading to the organization's failure to profit from new technology. This chapter has tried to illustrate the relationships between ILMD, SHRM, and selection and recruitment systems by further developing the typologies outlined in Chapter 4 and by illustrating how each type of firm uses recruitment and selection strategies based on the quadrant in which it is positioned.

Employment Strategies, Recruitment, and Selection 113

Once an organization has acted within its ILMD and SHRM constraints to hire an employee, it is faced with a further challenge — how can it assure the integration of the employee into the organization? The next chapter illustrates how context and structure interrelate with socialization and performance appraisal systems and how difficult it might be for an organization to obtain a fit between its structure and processes on the one hand and these SHRM subsystems on the other. It also highlights how a fit between the employee and the organization (micro-level fit), or a fit between processes and performance appraisal systems (organizational-level fit), can be obtained by considering the firm's choice in SHRM and ILMD. A combination of the insights associated with each of the perspectives in the previous chapters is used in an attempt to account for the emergence of organizations that effectively adopt technological innovation and survive the ensuing technology-induced change.

7. Socialization and Appraisal Systems

Once a new recruit has been chosen through the selection process discussed in the previous chapter, the firm will naturally wish to make certain that its planned investment in the individual pays off. To this end, it will try to recoup the expenses incurred in the recruitment/selection/training process by keeping him or her in the organization for as long as possible and by taking steps to guarantee a high level of performance from the employee. Although many individuals fit into the organization from the very beginning, and although the recruiting process itself may begin to mold others into compatibility, the organization often needs a specific plan to make sure that its employees remain well adapted. *Socialization systems* are, therefore, designed to acclimatize employees to the organizational environment — to help them "learn the ropes" and become comfortable within the organization — and *performance appraisals* are designed to give employees the feedback they need to maintain or improve the quality of their work.

Internal labour market determinism and strategic human resource management play a large part in determining the socialization and performance appraisal systems the firm will favour. Compatibility between these systems is a major concern, especially because making the appropriate choice enables the company to expand its choices in other areas, most notably in human resource management, just as making an appropriate choice in the recruitment and selection systems also has this effect.

We begin with a brief overview of the interrelationships between SHRM, ILMD, and organizational socialization and appraisal systems; following this, we turn to our typology once more and discuss how organizations in different quadrants approach socialization and appraisal systems.

Effects upon Socialization and Appraisal Systems

Socialization involves a systematic attempt to imbue the employee with the values and norms that characterize the organization, a process that is necessary to a certain extent if human resource strategies are to be effective and if the organization wishes to be successful in its endeavours. The ILM and management, which traditionally represent divergent views and values within the organization (operating almost like a two-party political system in this respect), usually disagree about what the socialization process should involve. From the perspective of management, the ideal socialization process would only give the employee the values that characterize management and thus result in an organizational machine composed of like-minded individuals who could operate with united efficiency. From the perspective of the internal labour market, however, a socialization process such as this would result in an organization of "yes-men" with a weak and ineffective ILM. A strong ILM, therefore, is likely to insist that the socialization process accomplish the objectives of both the management and the ILM, instructing employees in the values and norms that characterize both. The appraisal strategies that go along with the process would reflect this.

A look at Japan's Nissan Corporation amply demonstrates the relationships between socialization, SHRM, and ILMD. Even before Nissan opened its first North American plant in Smyrna, Tennessee, it realized that the plant would only be a success if the company could socialize its employees into an acceptance of Japanese business practices. Given that many of these practices were quite alien to Western workers, this task would be especially difficult; the company could not, therefore, afford the undermining influence of a powerful ILM. Hiring members of the American Automobile Workers Union (AAWU), it was believed, would have resulted in the creation of just such an ILM, because the union itself supported different values and beliefs that it would wish to protect and it could thus be expected to resist socialization. To avoid this potential problem, Nissan recruited a nonunionized work force composed of people who had never worked in car manufacturing plants before, ascribing to the theory that nonunionized employees might resist socialization less. Realizing, however, that its "foreign" philosophies would still be likely to cause some friction, the company sent all first-line supervisors to Japan for several weeks to become acquainted with the work roles of their colleagues in that country. Their supervisors' firsthand experience with the "Japanese" method of manufacturing cars and trucks made it easier for American employees to accept this method. Because the organization did not have to cater to the ILM, it was able to socialize employees quite successfully. We could not expect similar attempts to socialize the employees of Detroit-based car manufacturers into acceptance of the same procedures to be quite so successful.

Socialization

As the procedure through which firms bring new employees into their culture, *socialization* involves a number of processes: learning the rules, procedures, values, and norms of the work group, department, and organization; developing social and working relationships; and acquiring the skills and knowledge needed to perform the new job (Feldman, 1981). Socialization can be divided into three steps: (a) *anticipatory*, (b) *encounter*, and (c) *change and acquisition*. Anticipatory socialization encompasses all the learning that occurs before a new member joins an organization. The potential employee might talk to current employees, read about the firm in a magazine, or be provided with information about the company before a job interview. Assuming that this information gathering activity provides an encouraging view of the firm and the job, he or she might then begin to unconsciously accept the values of the organization — in essence, to socialize him- or herself. The encounter phase begins with the employee's first day on the job. He or she now learns what the organization is really like, and some initial impressions may change. In the change and acquisition phase, relatively long-lasting changes take place; the employee masters his or her job, adjusts to the work group's norms and values, and is successful in performing his or her new role. Most of the formal process of socialization takes place during the latter two phases, once the employee has been hired.

Feldman (1981) proposed a contingency model of socialization, which suggests that if the multiple levels of socialization described above are successful, there will be certain behavioural outcomes (e.g., the employee will remain with the organization) and affective outcomes (e.g., internal motivation). Thus successful socialization should lead to such behaviours as showing initiative, being punctual, and meeting deadlines. Affective outcomes could include job satisfaction and organizational commitment. Effective socialization can lead to a degree of cohesion among coworkers and their work groups that, for recently hired employees, may be one of the most attractive features of the job. Ineffective socialization can lead to low cohesion that may encourage employees to seek more attractive positions elsewhere. This latter fact appears to be confirmed by Sheridan (1985), who concluded that lower performance, absenteeism, and turnover reflect discontinuous responses of sociopsychological withdrawal by the employee. In its final stage, psychological withdrawal will result in the employee leaving the organization. Effective socialization should assure that the employee gains a clear understanding of his or her work roles and becomes part of the work group, thereby preventing any sociopsychological withdrawal symptoms from occurring.

Work Roles

The effective socialization of a new employee depends upon the role sets and behaviours he or she is supposed to acquire to perform successfully. In any context, a *role* is defined by its characteristic behaviours (Biddle, 1979, p. 58). The organization will quite naturally expect the individual to perform certain roles, and learning exactly what his or her superiors expect will enable the individual to fit in better and perform more effectively.

During each stage of the socialization process, the individual learns more about his or her work roles and the expectations of the organization and peers. In the anticipatory phase, the employee, in most cases, learns something about the roles he or she is expected to play. The organization might tell the potential recruit about the work role attached to a position during the recruiting stage, or later during the actual selection process; this may also serve to weed out employees who are indifferent to the position. In the encounter phase, the newcomer will learn more specifically what roles he or she is expected to perform and how to perform them. By the time the employee has reached the change and acquisition phase of socialization, he or she has become familiar enough with the expectations of his or her superiors to modify the roles that are part of the position. Certain roles not originally part of the position might become associated with the person filling the position at the time and thus, by association, with the position itself (Biddle, 1979, p. 65). One employee might, for example, take it upon him- or herself to make coffee every morning. When this employee leaves the firm, his or her replacement might be expected to do the same because the activity had now become part of the role. Newly created positions give the newcomer even more leverage in establishing his or her own set of role behaviours.

Interpreting Events on the Job

Newcomers wishing to learn their roles successfully should be encouraged to seek the help of veterans to interpret various situations. Veteran insiders normally know what to expect in a given situation. They are rarely surprised by anything that happens within the organization, and when something surprising does occur (such as a promotion being given to someone other than the expected party), they will usually have sufficient experience to interpret the event and make sense of it based on relevant knowledge of the immediate situation. The insider also has the advantage of having connections within the organization—other insiders with whom he or she can compare interpretations and perceptions (Louis, 1980). The new employee, in contrast, does not possess the knowledge and experience of the veteran, and although a precise job description and the impressions he or she develops of the organization during the recruiting and selection phase will give the employee some basis upon which to make judgments, there is likely to be a limit to how correct his or her interpretations of situations can be. He or she might, for example, interpret being passed over for promotion as a personal slight, and adjust his or her attitudes towards the employer accordingly; this would, no doubt, affect performance and socialization negatively. The insider, in contrast, might identify an altogether different motivation for the employer's actions that has nothing at all to do with the new employee personally.

Implementing Socialization

Implementation of socialization differs for employees new to the company and for current employees who move from one position to another. Changing jobs within the same firm requires only an encounter and acquisition/change phase, but new employees also go through anticipatory socialization. The reason for this difference in the socialization process is, of course, that current employees are already acquainted with the values and beliefs of the organization. Socialization for these individuals must simply prepare them for the new position but not the organizational environment in general.

New employees. Once the new employee has accepted a position, the organization may choose to begin the formal process of socialization by sending him or her a booklet before the first day of work that outlines organizational goals and philosophy, remuneration and fringe benefit packages, performance appraisal systems, career development methods, and other information designed to let the employee know what the company is about.

Facilitating the encounter and acquisition/change process for a new employee might be accomplished by use of a *mentor*, a specific senior employee to whom the rookie can turn for help. Using mentors is one method that career and adult development theorists have described as having great potential to enhance the development of individuals in both early and middle career stages, and studies of the mentoring relationship suggest that it can be instrumental in supporting both career advancement and personal growth of new employees (Kram, 1985).

Yet, even though the mentor may become especially important for "learning the ropes" and acquiring new skills, habits, and values, training may be crucial during the encounter stage. Training is intended to foster learning among new organizational members. Learning may be thought of as a process by which an individual's pattern of behaviour is altered in a manner that contributes to organizational and individual effectiveness. Hence, through organizationally sponsored training, the new employee might acquire some of the necessary job skills required to succeed in his or her new work role.

Current employees. It is natural to expect that the socialization process needs to be implemented only once, however, in actuality, employees may need to be "resocialized" occasionally. This is because a large number of people remain with the same organization for many years. Hall (1982) reported that over 40% of American men and 15% of women over 30 years of age spend 20 or more years with the same employer. Rosenbaum (1985) found that they will hold a variety of jobs within their organization during their careers. Transfers to new jobs may be unwelcome, however, and this, along with the possibility that the employee may eventually become bored with the organization, might require that employees' faith in organizational tenets be reaffirmed periodically. To do this, the organization should have a process for socializing long-term employees into new work roles.

Such a process might be facilitated by a variety of relationships within the organization. As we have already noted, it is easier to socialize individuals if they can be convinced to seek out the advice of their more experienced coworkers when they have problems. Even the current employee who gets transferred may need a specific senior employee (mentor) who facilitates his or her adjustment to the new position and required work role(s). A formal or informal mentoring system is not the only solution, however; other adult relationships in work settings, such as peer relationships, have also been shown to offer unique opportunities for personal and professional growth (Kram & Isabella, 1985).

Through strict job classifications and descriptions, or through the protests of workers and unions, ILMD limits the organization's ability to make changes in work roles and often causes the average employee to be socialized primarily to the values of the union rather than to the values of the firm. The organization trying to change employees' work roles may thus have little chance to socialize its workers. London's Fleet Street newspaper giants had to deal with this sort of problem in the mid-1980s when they tried to computerize newspaper typesetting. Typesetting jobs quite suddenly ceased to exist and former typesetters were given new duties. The typesetters, members of a powerful union, resisted such a dramatic change, feeling that management should have first consulted them and assured their willingness and cooperation. The result was a series of major strikes that crippled management's attempts to win worker acceptance.

Socialization and New Technology

The issue that must now be addressed is how socialization may affect or interrelate with the adoption of new technology and the ensuing technology-induced organizational change. During a person's organizational tenure, the socialization phase is probably the most critical stage for becoming aware of technology strategy and policy. For instance, some organizations may have a technology policy stating that new innovations and their adoption may not lead to layoffs of current employees. Instead, safeguards instituted by highly developed internal labour markets may prevent such negative outcomes (e.g., Betriebsrat, Max-Planck-Institut für Bildungsforschung, 1988).

Such a policy should come to the employee's attention during the anticipatory or at least the encounter phase of socialization. During the latter phase, the employee should also be made aware of the ways technology will be used in his or her work. Through social information processing, the individual tries to interpret the work situation by observing insiders working with computers or other technology within the social context of the job (Salancik & Pfeffer, 1977). In this way, the individual gains some normative understanding about technology and its potential effects in the new job situation.

The above indicates that beliefs held by new employees about technology, later use of new technology, and reactions towards technology-induced organizational change are most likely to be formed and set during socialization. Because research indicates that attitudes and beliefs are generally highly resistant towards change (Staw & Ross, 1985), it is very likely that those formed by the new employee during socialization will remain with the individual throughout his or her tenure with the firm. Thus poor strategy on behalf of management at this stage may cost the company a great deal over time.

Appraisal

Once the employee has completed the encounter phase — which is to say, once the employee has been socialized into the firm and work group and has learned to perform required work roles — his or her performance should be evaluated. The feedback provided by the performance appraisal lets the organization know whether the employee is inadequately socialized at present and suggests to both the organization and the employee the adjustments that might be required to make the third socialization phase more effective.

Level of performance is determined by measuring the appropriateness or desirability of the employee's behaviour within the organizational setting according to criteria based on either the behaviour of others or a standard established by management (Campbell, Dunnette, Lawler, & Weick, 1970). The *performance appraisal process*, therefore, involves the identification of these criteria, the actual measurement of performance against them, and the consequent review of the individual's performance. It is most important during the evaluation process to attempt to steer subsequent performance by developing new objectives and standards for the next appraisal period, a task in which the supervisor and employee often work together.

Performance is dependent upon previously established goals and the nature of these goals. Difficult and specific goals seem to have positive effects on performance (perhaps because they inspire the employee more). Research has shown that participative goal-setting causes subjects to select higher goals than a superior normally would (Locke, 1968), which seems to suggest that it is important to involve employees in the process of developing their goals. A *meta-analysis* (i.e., a family of techniques used to aggregate research evidence across studies) by Tubbs (1986) confirmed the positive effects of goal difficulty, goal specificity, and participation in the goal-setting process upon performance. Performance appraisal, because it provides feedback that can be used to set goals, is useful to that end. To be effective, appraisals must be done on a regular schedule and in systematic fashion, examining the employee's fulfillment of objectives for the period evaluated (Carroll & Schneier, 1982). However, if the objectives set at the beginning of the period are vague and do not include or address issues of quality, quantity and timeliness, it is difficult to evaluate actual performance and to use the process as a tool to help the employee improve his or her future performance (Locke, 1968).

Methods of Appraisal

Although individuals within the organization are usually appraised separately, the objectives they are given and against which their performance is measured must also be the objectives of their departments. Often, the employee may not only be assessed for individual performance but also for his or her work group's performance. A sales representative, for example, might be assessed for the performance of his or her colleagues within the same sales region. He or she could, therefore, improve his or her performance by helping coworkers improve theirs, and in the process increase the quality of his or her compensation, which would also be based on both group and individual performance.

There are a variety of performance appraisal methods and formats available to the organization, some of which we will briefly describe here (for an extensive review, see Bernardin & Beatty, 1984, chap. 4). In some firms, appraisals are done using a standardized form with specific sections addressing the performance of the individual in relation to his or her objectives. One of the more popular methods is *paired-comparisons*, which requires the appraiser to make relative comparisons between subject employees in terms of performance or organizational worth. The supervisor may compare all possible pairs of subordinates on their overall ability to do the job. Doing paired comparisons is a substantial task and the number of comparisons required can be calculated as follows—$N(N - 1)/2$ number of comparisons; N = number of employees. Thus if the supervisor has five subordinates, he or she will have to do 10 [i.e., $5(5 - 1)/2$] comparisons; —however, with 10 employees, the total number of comparisons required rises to 45! Such a performance appraisal process is, of course, very time-consuming.

Another approach to appraisal is *rank ordering*. The appraiser first selects the "best" person, then the "worst" person, then the second best, then the second worst, and so forth until all persons have been ranked. The *forced distribution* method requires the appraiser to "force" a designated portion of the subjects into one of several categories of performance. For instance, it could be that the appraiser must rank five individuals as performing above average, five at average, and the other five below average, much as, in education, a forced distribution (bell-shaped curve) is sometimes used by instructors who hand out a fixed percentage of As and Fs.

Rating scales, sometimes called *graphics* or *summated scales*, are appraisal methods in common use, particularly for positions below managerial levels. All rating scales call for judgment of rated performance levels along an unbroken continuum (e.g., excellent to unacceptable) or into discrete categories (e.g., superior, satisfactory, unsatisfactory) within a continuum. They differ in the types of aids or cues that are offered to the rater and placed along the continuum as well as in the number of distinctions along the scales (Carroll & Schneier, 1982, pp. 102-103). *The trait scale* is a popular approach that evaluates employees on the extent to which each possesses such traits as punctuality, dependability, friendliness, and ambition (Carroll & Schneier, 1982). The *critical incident* method consists of collecting reports of

Socialization and Appraisal Systems

behaviours considered "critical" in the sense that they make a difference to an employee's success or failure in a particular work situation and then assessing performance based on these using adjective-anchored (or graphic) rating scales.

Behaviourally anchored rating scales (BARS), the last rating scale to be discussed here, may be described as a graphic rating scale with specific behavioural descriptions at various points along each scale. One scale is used for each of the important aspects of job performance, including abilities, knowledge, skills, duties, responsibilities, or personal characteristics. BARS can be used to measure various components of job-related tasks such as quantity produced by the individual, the quality of one's output, attendance (timeliness, absenteeism, tardiness), and knowledge of standard procedures required to perform one's job (e.g., safety procedures). For instance, a nurse working in a cardiac unit would be expected to know certain emergency procedures commonly used in the unit and would receive a high ranking for methodically following the required steps when a cardiac arrest patient is delivered to the unit. If, however, the nurse became disoriented in an emergency situation, and was hesitant about exact procedures, his or her performance would be ranked below average. Thus the appraiser is given a scale with behavioural anchors (description of the behaviour associated with, for example, poor, below average, average, above average, and high performance rankings) when doing an evaluation of a subordinate. These examples of performance appraisal methods represent a fair sampling of the numerous approaches that can be used (see Bernardin & Beatty, 1984, chap. 4).

Source and Accuracy of Data

The data required for performance appraisal can be obtained from various sources. Most often the appraisal is done by the subordinate's immediate supervisor; however, the organization might also require the subordinate to do a self-assessment or even use his or her peers and subordinates as additional sources of data. Including the subject in the appraisal process allows the employee and the supervisor to compare their interpretations of events. Using the perspectives of colleagues—some of whom are directly affected by the individual's performance and some of whom are not—can augment the data's *validity* (the extent to which an appraisal method/instrument measures what it is designed to measure) and *reliability* (the extent to which an appraisal method/instrument provides consistent results), thereby providing a more accurate picture of the employee's actual performance and a control that may help to expose extremely biased ratings. Biases in appraisals apparently do exist. Some research indicates that females are often less highly evaluated than their male peers (e.g., Dobbins, 1985), and other research has shown that subordinates with whom supervisors have established relatively high task relationships and personal acquaintances receive significantly more favourable overall performance ratings and are more likely to be promoted than other subordinates (Kingstrom & Mainstone, 1985).

Researchers have long been interested in identifying methods of improving the quality and accuracy of performance ratings. The key concept towards understanding the cognitive processes of the individual performing the evaluation is called *categorization*. Individuals perceive and process information in terms of abstract categories defined either by various schemata or, more rarely, by somewhat more concrete prototypes. These categorization blueprints, which may be based on formal or informal sources of information, allow individuals to achieve "cognitive economy" by reducing the amount of information processed and stored (e.g., Behling, Gifford, & Tolliver, 1980). If a behaviour, therefore, is congruent with the expectations defined by the categorization blueprint, it is noted and categorized automatically — a consistent mapping condition. If the observed behaviour is *inconsistent* with categorization schemata or prototypical expectations, however, conscious attention must be used to recategorize the individual's behaviour — a variable mapping condition (Ilgen & Feldman, 1983). Once the behaviour of the individual has been recategorized, further perception and recall of the individual's behaviour in future situations will then be biased towards the new category. As a result of categorization, performance appraisal ratings are thus more accurate when the behaviours of the evaluated party are consistent with the expectations of the evaluator (Mount & Thompson, 1987).

The categorization theory has been confirmed in research showing that evaluators are biased in favour of recalling behaviours that are consistent with their general impression of an evaluee (Murphy, Gannett, Herr, & Chen, 1986). Yet, although the subject employee's subsequent performance may systematically alter the evaluator's recall of the employee's previous behaviour, his or her chance for a fair and accurate performance appraisal may be affected by general beliefs that have nothing whatsoever to do with him or her personally, even if performance improves. For example, Waldman and Avolio (1986), using meta-analysis, reported a pattern of increases in performance (as measured by productivity indices) at higher ages. Supervisory ratings, on the other hand, showed a slight tendency to rate older employees lower, which suggests that supervisors might automatically place older employees into categories that are unrealistic — assuming perhaps that older employees are likely to tire easily, "lose their competitive edge," and so on. Performance ratings also showed more positive relations with age for professionals as compared with nonprofessionals — again, possibly the result of a categorization system that classifies older professionals as "wise grey heads" and older nonprofessionals as "tired old men." Given the numerous factors that may influence the accuracy of performance appraisal ratings, the best way to calibrate the appraisal scales may be to use as many evaluators as is possible and economically feasible.

Use of Data

The organization's use of the data it receives through performance appraisals will reflect the particular conditions of the quadrant in which it is situated. For instance, in Quadrant C, where bonuses and rewards make up an important portion of the employee's compensation package, appraisal ratings are used to determine the magnitude of these bonuses and rewards. In Quadrant A, however, where bonuses and rewards are less important, the usefulness of appraisal data is reduced. When there is no link between performance and any kind of reward, and when future employment or promotions are based on such things as seniority rights rather than performance, the performance appraisal is often not worth the time the organization may invest in it.

Performance Appraisal and New Technology

A comparison of the use of "older" technologies (e.g., typewriters) with today's more sophisticated equipment (e.g., computers) indicates that investment for each single position, including socialization and training costs, has increased many times during the last several decades when calculated using constant dollars (i.e., adjusted for inflation). Today's worker will use technology that may be valued at multiples of thousands of dollars. For this to be worthwhile, the employee must use the new technology effectively.

The performance appraisal can be used to provide the employee with feedback about how he or she is performing in general job dimensions (e.g., job knowledge, interpersonal and leadership skills) and also in the *technology dimension*. To make technology-induced organizational change a complete success, it is necessary that work is organized to give the individual the command over job tasks needed to become a high performer. Based on Table 4.2 (in Chapter 4), it is obvious that the four organizational types may have to handle performance appraisals differently, but in all quadrants, performance appraisals may help to identify problem areas, such as work organization and authority structures, that may evolve due to the introduction of new technology. Subsequent organizational adjustments are needed to enable individuals and groups to perform at their peak.

Another important issue is that, because appraisal data can be used to advance the employee's career development within a corporate hierarchy (Rosenbaum, 1985) by identifying strengths and weaknesses and by indicating where additional training is most needed, an extensive appraisal data base might make it easier for the firm to identify the individuals most qualified for future job openings within the corporation. Career development, therefore, will be far more effective if the appraisal system is accurate, systematic, and regular.

Organizational Type and Socialization and Appraisal Systems

Following the pattern established in previous chapters, we now examine the relationships between ILMD, SHRM, and socialization and performance appraisal systems by looking at several of these systems in terms of the four by-now-familiar quadrants.

Socialization

Overall, the few systems that organizations in Quadrant A have for socializing employees are informal and unstructured. Firm-initiated anticipatory socialization is rare, and the encounter phase may be influenced by the ILMD in such a way that a number of new employees may spend the first day on the job together, being introduced to the company's values, fringe benefit packages, and so forth by tenured colleagues and personnel representatives. The acquisition and change phase is, however, not structured but occurs on a more informal basis "on the job"; colleagues, the work group, and possibly the union steward fill the gaps in a newcomer's knowledge about the firm on an informal basis and also teach him or her the skills needed to perform well on the new job. An example might be an employer like Canada Post, which has no clear strategy for socializing its employees except during the anticipatory phase. Hence, the information given during recruiting and selection is somewhat determined by the ILM (e.g., job descriptions and duties, type of testing if any, and interview procedures), but, once the employee has started in the new position, the encounter and the change/acquisition phases are left to colleagues and the work group. The organization

TABLE 7.1: Four Human Resource Strategy and Internal Labour Market Types: Socialization, Appraisal, and Career Development Systems

Variable	Quadrant A High ILM Determinism, Low Choice in Strategic Human Resource Management	Quadrant B High ILM Determinism, High Choice in Strategic Human Resource Management	Quadrant C Low ILM Determinism, High Choice in Strategic Human Resource Management	Quadrant D Low ILM Determinism, Low Choice in Strategic Human Resource Management
Socialization	Informal/unstructured	Formal	Formal approach, mentor or peer is often designated	Depends on the department
anticipatory	Defined according to ILM	All three steps are used and are identifiable processes	All three steps are used representing three distinct processes	Not done or left to recruiter's discretion
encounter	Left to work group or union steward			Depends on supervisors and work group
change and acquisition	Left to work group			Possibly done by peers and work group, informal
Appraisal	Impersonal: based on return on investment, productivity, and subjective assessment of contribution to firm; outcome is job classification related	Systematic: goal setting groups and/or job classification based, focusing on quantity, quality, timeliness, and costs Source of data: superior, self, and subordinate	Systematic: goal setting individual based focusing on quantity, quality, timeliness, and costs Source of data: superior, self, peers, and subordinate	Interpersonal and subjective: allocated in a paternalistic manner, based on supervisor and/or division head Apathetic and unsystematic
Primary Focus	Group	Individual and group	Individual and group	None in particular

tends to ignore such opportunities as training or skill upgrading to help job holders either acquire the skills necessary to perform well or learn new skills to manage technological change.

In contrast, the Quadrant B firm has distinct plans for using each of the three phases. For instance, a hospitality night on a university campus might be designed to provide anticipatory socialization, encounter socialization might involve having most new employees spend their first day on the job according to a plan given them at least a week in advance. Change and acquisition socialization is usually formalized by providing an employee with the necessary training to succeed in his or her job role. Siemens, the German electronics multinational, is a good example here. Information concerning the company (e.g., the last annual report and brochures about the division in which the person is applying for a job) is provided to the applicant before the job interview. Then, on the first day, new employees are shown around, introduced to their peers, and allowed to talk to future coworkers. In-house training provides the skills the employee needs to prepare for new assignments and work role changes due to technological developments.

The firm in Quadrant C is very similar in its approach except that a mentor or an experienced peer may be assigned to a newcomer to help him or her get acquainted with the job and to interpret events happening in the work environment that are difficult for outsiders to understand. Arthur Andersen & Co.'s consulting group is a good example here. Shortly after joining the firm, each employee is sent to Chicago or Geneva for a three-week seminar to acquire the necessary job skills and to be familiarized with the company's values, expectations, and employee benefits. After successfully completing this training seminar, the newcomer is assigned to a project team in his or her home office. The third phase of his or her socialization is facilitated by frequent interaction with the team leader on a formal and informal basis. Managers will usually perform some mentoring functions for the junior members, functions that are mostly informal but that are expected and strongly encouraged.

The decentralized nature of Quadrant D firms means that socialization strategies are determined by individual departments or by the recruiting party. Most academics interested in working at a North American university (which, in this context, is characteristic of this kind of firm), therefore, experience a degree of anticipatory socialization that depends upon the recruiter's perception of the job's importance. Brochures and resumés of potential colleagues might be supplied to a candidate seeking a position in, say, the management department, but little material might be provided to a different applicant wishing to join, say, the physics department, which might have a different attitude towards recruiting. Later socialization will be informal and depend on the supervisor and colleagues.

Appraisal

In Quadrants B and C, appraisal of the individual's performance often involves self-appraisal as well as the more common evaluation by superiors; on occasion, even the opinions of subordinates may be sought. In this context, the most important difference between these two quadrants is that organizations in Quadrant B are likely to favour group- and/or job classification-based appraisals, but those in Quadrant C are more likely to favour individual-based appraisals.

A firm in Quadrant B, such as Siemens, would appraise its electricians working on a power plant construction site by obtaining data from the individual and the supervisor. The employee is also affected by the performance of others (e.g., completing a phase of an overall job using fewer hours than budgeted by Siemens that had already been paid for by the customer) because a portion of his or her bonus is affected by the group's performance (that is, if 90% of allocated hours are used, the saved portion of 10% is paid out as a bonus to the site crew).

The situation in Quadrant C may be illustrated by the example of a Procter & Gamble sales representative who is required to perform self-assessment and who is typically appraised not only according to his or her own performance but also according to that of colleagues within the same region. This appraisal is immediately followed by systematic goal-setting, which will be characterized by a clear focus on quantity, quality, time and costs. Such a focus is natural for this quadrant, given that innovation here (as shown in Table 3.1, Chapter 3) is productivity-driven.

Organizations in Quadrants A and D do not always adopt performance appraisals as assessment tools, although not necessarily for the same reasons. High ILMD in Quadrant A should force firms to be more systematic in their approach to performance appraisal, yet appraisal nevertheless remains primarily based on productivity and on subjective contributions to the firm. This may increase worker grievances and thus decrease profitability substantially (Ichinowski, 1986).

To illustrate this somewhat, although Canada Post uses appraisal systems, benefits for the employee depend very much on the overall financial situation of the organization (which in this case was striving to achieve deficit reduction, profitability, and better service by the end of 1990 — price hikes accomplished the former two and failed with the latter). Because the effect of appraisal upon the employee's compensation package depends on this financial situation, the employee's actual job classification, and the union contract, a whole group of employees with the same job classification may get similar performance rewards. This equalization procedure defeats the ultimate goal of performance rewards, making them more like cost-of-living adjustments. Quadrant D organizations, on the other hand, may not use appraisals because they have not yet developed the skills and strengths to administer a valid and reliable performance appraisal system. The example of the Swiss PTT will serve here. Although most letter carriers, telephone operators, and field-workers do receive performance appraisals, these are usually subjective and based on traits, and their use and focus depend largely on the supervisor and/or division head.

Socialization and Appraisal Systems: Facilitating Technology-Induced Organizational Change

The previous section illustrated how various types of organizations may use socialization and appraisal systems depending upon ILMD and choice in SHRM. Technology-induced organizational change does, however, require some additional effort to make socialization of new employees a success and make the appraisal system more effective, thereby helping the individual to attain higher performance levels. Most importantly, during socialization the new employee must be made aware of the company's technology policy and given the company-specific skills needed to use the technology. As the next chapter will discuss, training systems may play an important part even during the encounter phase of socialization.

When looking at performance appraisals and new technology, it may become necessary to increase the focus of the appraisal from a singular level of analysis (e.g., individual) to at least two levels (e.g., group and individual) as done by Quadrant B and C firms. Research indicates that increased use of advanced technology and equipment throughout an organization makes most people's performance levels more and more dependent upon cooperation with, and the performance of, other individuals (e.g., Bikson, Gutek, & Mankin, 1987). Hence, the working group is becoming more and more central to the performance of the individual.

The technology dimension must also be addressed by performance appraisals. Because the investment for each employee's work technology is increasing rapidly, factors that affect technology-related performance should be included in an appraisal. Areas of interest and possible examination may be the extensiveness of technology use (e.g., uses the typewriter instead of the word processor), know-how (e.g., uses technology without much help from others), security (e.g., keeps data secure), and productivity. Rewards based on technology-related performance can then be paid, serving as motivation for the increasingly effective use of technology in the future (see Chapter 9).

Inevitably, a high ILMD may make it more difficult for a firm to integrate the technology dimension into the performance appraisal without consent from its work force (e.g., union). Once again, an appraisal system with a technology dimension requires systematic goal setting and an approach that is more likely to occur in Quadrants B and C than in A and D. This is yet another motivation for the latter two types of firms to try to change their positions and move into Quadrants B and C, respectively.

Summary

Although there are a variety of approaches and techniques available to the organization wishing to develop effective socialization and performance appraisal systems, the specific amount of internal labour market determinism and choice in strategic human resource management faced by the organization determines which the firm will use or is capable of using. In organizations with greater choice in SHRM, socialization and appraisal systems are formalized and systemized; in organizations with less choice in SHRM, they are often not used at all. Knowing what we do, we can, as an example, safely predict that an organization in Quadrant A that tries to use socialization techniques like those usually applied in organizations in Quadrant C will encounter a host of difficulties. First, it does not have the freedom to manage the complicated management strategies of socialization and appraisal that are used in the latter quadrant, and, second, its union-dictated ILM makes it extremely difficult for the company to establish the kind of rapport with the common worker that effective socialization requires.

A company's approach to recruitment and selection necessarily affects the characteristics of its socialization and performance appraisal systems. The company in Quadrant B, for example, will use systematic criteria based on job descriptions and so forth in the recruiting and selection process. Already, this formalizes the anticipatory socialization phase to some degree. Applicants with the necessary skills are encouraged to believe that they "have what it takes" to be successful within the organization if they already know what is expected of them and thus begin their socialization. An applicant to an organization that had advertised a position with only a three-line newspaper advertisement will not know much about his or her possible duties within the organization and cannot begin to be socialized until he or she finds out what the job really entails.

The socialization and appraisal strategies presented in Table 7.1 fit the quadrant in which the firm is located when looking at the level of ILMD and level of choice in SHRM. Innovations may, however, require changes in approach. For instance, as discussed earlier, socialization should prepare the individual for the work environment and technology-mediated work, and the appraisal should help identify the areas where structural changes (e.g., work, communication, and negotiation systems) are needed to achieve a fit between new technology (or innovation) and the organizational characteristics (external and internal, see Table 4.2), thereby improving individual- and group-level performance.

8. Career Development and Training Systems

Career development and training systems are affected by internal labour market determinism and choice in strategic human resource management in much the same way as the recruiting, selection, socialization, and appraisal systems discussed in the previous chapter. The organization with a range of choice in SHRM policies, for example, is more likely to have a work force that is well socialized, which in turn is important to career development and training because the degree of the employee's socialization determines, to a great extent, his or her expectations regarding career development. The organization with a low level of choice in SHRM, on the other hand, often has a tough time implementing the kinds of career development and training strategies it wishes to use, especially if it is also subject to a high level of ILMD. ILMD can limit or even dictate the organization's approaches to training and career development. Industrywide union agreements, for example, can discourage novel approaches to career development problems by demanding the same methods of all employers. This is especially likely to be the case in companies that do not possess comprehensive career development systems and are forced to deal with career development issues as best they can whenever they arise. Union contracts may also insist that training be based on job classifications, seniority privileges, or demarcation lines (e.g., union membership) rather than on the individual's need and potential.

This is not to say that union contracts are entirely negative in their influence; from the employee's point of view, the ILM often aids career development and training by demanding retraining strategies, to offset the effects of innovations, that are more effective than those proposed by management (see IG-Metall, 1984). In such a case, ILMD may even make a career change easier. However, for the employer and for the particularly ambitious employee, the end result of an overly powerful ILM is usually the same: highly structured training systems that are too dependent upon union contracts and leave little space for individualized career development plans (Windolf, 1986).

As argued in Chapter 3 (see Figure 3.1), choice and determinism are not at opposite ends of a single continuum of effect but in reality represent two independent variables whose interaction or interdependence must be studied to explain technology adoption and technology-induced organizational change. Both training and career development strategies are affected by this interaction. Environmental determinism (e.g., market deregulation) may force the firm to adopt new technology to remain competitive, thus necessitating training. Bringing another facet of environmental determinism into play, however, it is possible that the firm can use government-sponsored programs to train its employees for the new technology, thereby offsetting some of the costs associated with its introduction. Having the degree of choice needed to utilize these external programs reflects directly on the extent to which they are exploited in a career development system. Summed up concisely, the chain goes like this: Career development and training systems must be coordinated with a company's technology transition strategy, which in turn is affected by the firm's strategic choice and environmental determinism.

Career Development

For most managers and organizations, *career* has traditionally been defined as a *linear* progression up the corporate ladder. According to this definition, *mobility* (measurable by frequency of promotions) is essential to a successful career, but remaining in one position for a long period, in contrast, is seen as a sign of stagnation (Rosenbaum, 1984). This was confirmed by Kotter (1982), who surveyed a group of executive managers in a large corporation and found that those headed for the upper echelons had usually been promoted out of their previous positions within 2.4 years. Given societal and demographic trends, however, it appears that an upward linear progression will become less characteristic of career success in the future. Baby-boomers saturating executive positions and the movement away from early retirement will contribute towards making upper-management positions relatively inaccessible (Driver, 1985). Slow economic growth, meanwhile, can also be expected to reduce the number of new openings.

Though the dwindling opportunities for advancement may mean that *career* will be differently defined in the future than it is now, there is no reason to suspect that this will have catastrophic effects. Career *satisfaction*, it is important to remember, has never been entirely dependent on anything so simple as hierarchic rank. Employees are also interested in opportunities for career and personal growth, which suggests that lateral mobility may be important as well. In this context, *lateral mobility* could be defined as a change in the type of work one does (e.g., a move from marketing to sales or accounting) without necessarily changing one's rank in the hierarchy (Schein, 1978). Gattiker and Larwood (1986a) found that *both managerial and nonmanagerial* employees desire jobs that are interesting and that provide them with opportunities for decision making and communicating. The presence of such job characteristics can make the employee willing to accept less financial and hierarchic success than otherwise (Gattiker, 1985a). If, on the other hand, employees perceive few opportunities for growth, their commitment to the organization is likely to be reduced (Gattiker & Larwood, 1988; Landau & Hammer, 1986). Individual perceptions will vary according to the relative significance the employee attaches to the various career attributes; the importance of these perceptions makes an understanding of them vital to the career development system.

Tiedeman and O'Hara (1963, p. 46) defined *career development* as personal development viewed in relation to career choice and entry as well as progress in educational and vocational pursuits. From the organizational perspective, *career development* may be defined as the attempt to make the individual's tenure with the organization as successful as possible by employing him or her in a capacity that makes use of his or her talents. Thus both growth needs (i.e., opportunities for self-direction, learning, and personal accomplishment at work) and other needs (e.g., lateral movement, salary, and responsibility) are satisfied with successful career development (Hackman & Oldham, 1980, chap. 4).

Career development involves activities in which individuals participate to improve themselves relative to their current or planned work roles as well as activities sponsored by the firm to assure that it will meet or exceed its future human resource requirements. A *career development system* is used to coordinate these activities. This system is an organized, planned, ongoing effort to achieve a balance between the individual's career needs and those of the organization's work force (Leibowitz, Farren, & Kaye, 1986, p. 4). As such, it integrates the activities of employees with the policies and procedures of the firm and facilitates the development and refinement of present human resource activities.

Strategic contingency theory and resource dependence have important implications for career development. Because the resource-dependence model suggests that individuals whose performance is more important to the organization's well-being will be treated differently from others, it is not hard to conclude that they will be offered more career opportunities than less pivotal staff, especially if their skills are in great demand on the external labour market (e.g., Driver, 1979). Neither is it difficult to realize, following the resource-dependence model, that prominent positions within the organization usually offer more career development to their incumbents than other positions that are less visible. Approaches such as these may lead to friction within the ILM, primarily because they violate the equal wages for equal work principle. Court cases in the United States have not, however, been very conclusive on this matter (Cooper & Barrett, 1984).

An example of a career development system is IBM's Employee Development Planning System (EDPS), which allows the individual to do a self-assessment and career path exploration, define his or her objective and development needs, and (most importantly) prepare him- or herself for the manager-employee planning meeting (Minor, 1986a). EDPS, in turn, enables IBM to communicate the changing skill, job, and career environment to its employees. The system consists of three phases: In the orientation phase, the employee evaluates planning readiness by gaining self-knowledge and becoming aware of jobs; in the self-assessment phase, he or she creates a self-description profile including data on work interests, abilities, skills, and experience needs; in the final phase, the system matches the employee profile with job profiles using a job list (Minor, 1986a), and the employee reacts to potential jobs according to the responsibility they will entail, the skills they will require, his or her own willingness to relocate, and the overall appeal of the position. The system summarizes trends in employee reactions and the employee summarizes perceived development needs. The results are then reviewed with the manager (Minor, 1986b).

The EDPS approach is thorough, extensive, and highly individualized. It is also exceptional in that it is in an advanced stage of implementation and is designed to be used by employees from various levels, thereby encouraging them to become involved in their own career planning and development. In many organizations, however, career development systems do not really exist, primarily because *centralized data bases*—which are complex, expensive, and time-consuming—have not been created. To demonstrate, in 1983, Credit Suisse initiated a career development system designed to be implemented in five phases. It was not until the middle of 1988 that the company completed the third phase, at which time the data base included, for each employee, all pertinent information relating to salary, compensation, education, training (received both before and during Credit Suisse employment), and performance appraisals (see Hefti, 1983). With the completion of the fourth phase, the organization will be able to make use of the material within this data base for the purposes of human resource and career development planning. However, the system is still far from complete, and its finalization and implementation might not occur until the beginning of the 1990s (Zuercher & Mueller, 1986).

Career Development Systems and New Technology

A career development system can be extremely important in managing human resource problems created by technology-induced organizational change because it can determine change-related training needs and see that these needs are met. As we saw in the previous chapter, the major newspapers of London's Fleet Street did not plan for the rapid changes brought about by installing computer technology in the typesetting rooms — a strategic error that led to a considerable degree of labour unrest. In contrast, *Tages-Anzeiger* (*Daily News*), a major Swiss newspaper, carefully planned to adapt to computerization over a 10-year period and used its career development system to modify its work force accordingly. New typesetters were not hired when old ones left, and veterans were told about the 10-year plan and given choices ranging from early retirement to retraining to finding employment elsewhere. Most typesetters chose to retrain, and the time frame made it possible for them to find suitable work within the newspaper when their retraining was complete — an activity that would have been impossible for an organization without efficient career development and training systems.

Nonetheless, merely having a technology transition plan may not encourage employees to make timely career-related decisions, even when it is known that change is necessary. For instance, an automation pact was negotiated in 1974 between New York City's daily newspapers and their unionized printers, thereby ending the printers' decade-long veto of revolutionary processes in typesetting and composition that had been advocated by publishers. An agreement was reached by giving both sides full protection on the issues that each considered paramount (for management — installation of new technology, flexibility in assigning personnel, and abolition of wasteful work rules that had been sacrosanct for a half century; for printers — lifetime job security, emoluments, and safeguards of a kind not enjoyed by any of the papers' other employees; Rogers & Friedman, 1980). Printers, however, did not start to make career-related decisions (choosing between generous early retirement or severance packages or retraining for computer-mediated work) until a late stage during the three-year transition.

Because career development had never existed for the printers before — most of them had been printers for all their working lives — it was unreasonable to expect them to comprehend and participate effectively in career development overnight; adjustment always takes time. If career development systems are a permanent fixture of the organization, however, employees will learn to make use of them on a continuous basis. Such an approach allows the employee to assess his or her career opportunities with the organization, easing technology-induced changes required from the worker.

Training

Training in an organizational context involves any organizationally initiated procedures intended to foster learning among organizational members. *Learning*, in the same context, may be thought of as a process by which an individual's pattern of behaviour is altered in a direction that contributes to organizational effectiveness (Gattiker, 1990a; Hinrichs, 1976) — in this it bears similarities to the socialization process discussed in the previous chapter. As the preceding example of the Swiss newspaper indicates, training can be a valuable component of any firm's HRM strategy for managing technological change. Training should ideally provide managerial employees with generalized problem-solving and decision-making skills applicable to a wide range of problems that might be encountered within the organization (Campbell, Dunnette, Lawler, & Weick, 1970). To this end, the firm should also attempt to provide employees with (at the very least) a basic knowledge of word processing, computer-aided statistical analysis, spreadsheet and data base management, computer languages, and programming because organizational research has suggested that these are important areas in which managers should have skills (Jones & Lavelli, 1986; Taylor, 1980). Yet, because even lower level employees appear to use complex machinery more successfully when they understand the principles behind it, training is more worthwhile the more sophisticated its content; narrow skills training in itself may be insufficient (Tornatzky, 1986). While penny-pinching employers might argue against the cost of providing extensive training for lower-level employees, costs can be distributed over a longer period of time than for managerial employees. Because in both Europe and North America people above 35 years of age tend to stay more than one decade with the same firm, any argument that suggests training for nonmanagerial employees is impractical from a return on investment point of view is weak (e.g., Carter, 1988; Gattiker, 1988a).

One of the most traditional training methods is the lecture approach, which usually involves a carefully prepared oral presentation on a subject by a qualified individual (Reith, 1976) and which is generally conceptually and theoretically focused. Drill-and-practice, or "hands-on" training, conversely, emphasizes the application of concepts and theories to the solving of problems in a possible work setting. The learning mechanisms involved in this teaching method are *association* (built up through contiguity, which is established via practice) and *imitation* (through which learned responses become learned habits; Thorndike, 1913). Assignments done on the student's own time give the necessary practice for an acceptable level of contiguity, and exams help the student to learn how to perform under time pressure. Imitation may be accomplished by having the student observe the instructor while he or she performs certain tasks. Feedback allows for the modification of the student's behaviour and acts as a reinforcer (Bandura, 1977).

The training process should, as Burke and Day (1986) note, make use of several methods to teach skills. Using only one method often leads to gaps in the trainee's education: lecture training cannot provide the kind of experience that hands-on training provides, hands-on training cannot provide the kind of theoretical background that lecture training provides, and so on. It is not surprising, therefore, that employees who are expected to teach themselves only with the aid of a manual — a fairly common training method — are usually dissatisfied with the end result (Bikson & Gutek, 1983). Existing organizational training strategies typically use one or a combination of the methods described above, though the relative effectiveness of these in improving the acquisition of job-related skills remains questionable (a situation reflected by the literature on the effectiveness of training, which is full of conflicting results and unanswered questions; see also Chapter 12).

An increasing number of industries are dependent on innovation, especially in fields where knowledge changes rapidly, and the growing popularity of computerized technology means that almost every worker will have to have some degree of computer literacy to be successful in the workplace (Leontief & Duchin, 1986, chaps. 3, & 4). This kind of rapid change in the work environment has altered many common perceptions concerning the formal education and training of the work force. Traditionally, workers have been expected to obtain most of their education and formal training in their youth, prior to entering the labour force. The company hiring an accountant, for example, could be relatively certain that the employee would have the appropriate skills, and "training" would amount to the "fine-tuning" of these skills to the specific needs of the organization. Only a few persons in relatively high-level positions, therefore, could expect to undergo further in-depth education and training during the remainder of their work lives. However, the new demands of innovation have changed the status quo and have created a need for *recurrent comprehensive training*, even for not so "youthful" employees, between the ages of 35 and 55, in the prime of their careers (Leontief & Duchin, 1986, chap. 4). This training must alternate with work activities and is essential to many workers from all levels if they are to remain productive in their disciplines and, in some cases, if they wish to renew their licenses. Innovation has made recurrent training a necessity for unemployed people as well, given that they, especially, cannot afford to neglect upgrading their skills (and government-sponsored programs often give them the opportunity to upgrade; see OECD, 1973, p. 24).

Employed individuals may undergo recurrent training at the request or demand of their employers or simply out of a desire to better themselves. This training may be provided by the employer or by another organization contracted or chosen by the employer, such as a private firm specializing in skills training or a postsecondary institution. Analysis of the recurrent training policies (usually called "education policies" by governments) characteristic of various countries suggests that employees' motives for retraining, and organizational approaches towards it, may be dependent upon laws, union contracts, and other environmental characteristics. In the United States, retraining thus typically involves employer-sponsored programs, but in some European countries and Canada, costs are more likely to be government subsidized (see Levin & Schuetze, 1983).

In Sweden, the Educational Leave Act of 1975 guarantees the right of all public and private employees to take leave for educational purposes during working hours. To qualify, an employee must have worked for his or her company for 6 consecutive months or for a total of 12 months during the last two years (though this minimum employment qualification does not apply to training that is trade-union-initiated). The right to educational training covers all types of education — general, vocational, and union supplied. The act does not stipulate the maximum or minimum duration of leaves, nor does it say how many employees from a company may be absent at the same time; these points are settled by negotiation involving representatives of the employer and the work force (Rubenson, 1983).

Although Swedish workers are not paid for the time they are absent, they can apply for financial support from the government. Given such an environment, it is not difficult to speculate that retraining approaches involving company-originated training programs are less common. In most countries, however, governments have nothing to do with regulating educational leaves, and organizations and unions must hammer out agreements relating to this subject on their own. In the United States, for example, lower-level managers may be encouraged to attend managerial courses at a local university on their own time, with reimbursement by the employer for tuition and book fees upon successful completion of the courses. Some companies may even pay tuition and book expenses for courses of a more general nature. In such a setting, where the organization is forced to pick up much of the cost of retraining the individual anyway, the alternative of initiating specific company retraining programs, whether extra- or intraorganizational, would become more plausible.

Extraorganizational Training

Training and retraining outside of the organizational setting often involves highly specialized and advanced subjects and, for this reason, postsecondary institutions are usually expected to provide it. Such institutions have experienced a recent boom in the demand for executive training programs providing advanced training for a variety of industries in such areas as marketing, leadership, and production management. In 1987, the Harvard Business School's income from such programs represented 85% of the tuition fees collected in their much larger MBA program, a percentage that is still growing. Generally these fees were paid by the manager-student's employer, who usually also paid for all accommodation and travel expenses. These lucrative executive programs are often taught by instructors who may also have performed consulting jobs for such organizations. The content of the courses is thus usually enriched by in-the-field experience and is, therefore, deemed adequate for the organization's needs.

Although most North American postsecondary institutions have been fairly quick to respond to the demand for these kinds of training programs, they have been somewhat sluggish in reacting to the increasing demand for recurrent training from lower-level managerial and support staff, which involves a far greater percentage of the work force (Center for Public Resources, 1982). Response to the need has basically been limited to offering courses in the evening that allow part-time students to pursue their studies while working full-time, and to offering noncredit or study-program-affiliated courses through continuous education departments.

European institutions have done little better in meeting a corresponding demand for recurrent training. Though most European governments provide some kind of support for individuals who need to change their careers due to technological change or health reasons, part-time students are almost as rare as Fabergé eggs. Continuing education programs at the university level are practically nonexistent (the exception being university programs for seniors, accessible to retired individuals who may not meet normal university entrance requirements, which is not of great use in supplying the kind of recurrent training we are concerned with here).

In an effort to meet some of the demands placed on them, several North American universities have in recent years turned to private companies that provide skills training in the form of hands-on exercises and role-playing in the belief that skills training facilitates the transfer of theoretical knowledge in an applied setting. Development Dimensions International (DDI) is one such private firm that offers these services to business schools across North America (DDI, 1984). However, private companies such as this are normally used to provide recurrent training of a variety that is useful only to managerial personnel. And although they make recurrent training available as low in the hierarchy as line supervisors, they are usually not up to the task of providing training relevant to the needs of support staff.

Part of the reason for the lack of recurrent training programs in most of the Western world is that Western culture has traditionally assumed education is for youths; work is for young and middle-aged adults; and neither is appropriate for senior citizens. The location, hours of instruction, and structure of educational programs reflect this assumption. Faced with the prospect that outside institutions may be of little benefit in equipping workers with the skills they need to adapt to innovation, the company may decide to use in-house training instead.

In-House Training

Internal training can be more effective than external training because it can provide, in addition to job-specific and skill-oriented training, socialization-related training that gives instruction in the philosophy and values of the company and shows how these apply to such things as customer service, supervisory activities, and so forth. For example, acquiring computer skills in-house provides the firm with the opportunity not only to teach the individual how to use the new technology effectively but, equally significant, to instruct the employee about the organization's procedures and rules concerning such functions as copying or purchasing software, data security, and information retrieval. External training would give the employee some generic information about data security but not on how his or her employer specifically deals with these issues.

Internal training can, therefore, stress training in a mode of *thinking* as well as training in a mode of *doing*. External training, in contrast, is less likely to give the trainee a sense of the organization's values (being more likely to make the employee sympathetic to the values of the instructor) unless the external course has been specifically chosen by the organization with this in mind. The seminars used by Arthur Andersen & Co. described in the last chapter, which are designed to infuse new employees in the management consulting division with the company's values and philosophy, demonstrate how a company can take advantage of in-house training.

Because internal training instills mainly company-related "know-how" in the individual, skills acquired in such training may be quite job- and company-specific. Hence, such knowledge may not be easily transferred to another employer, thereby limiting the potential for other firms to poach valuable workers (Feuer, Glick, & Desai, 1987). Chapter 9 discusses and illustrates how such company-specific know-how may also allow the firm to offer lower compensation packages to its workers.

Young employee training. The specific content of in-house training varies from group to group within the firm; younger employees in particular will be likely to have quite specific needs. According to one U.S. survey, 75% of American firms claimed they had to provide many of their younger employees with remedial training programs in many basic skills that theoretically should have been acquired in school. AT&T, for example, spends more than $6 million annually to train about 14,000 employees in basic reading and mathematics skills (Center for Public Resources, 1982). In another survey, 35% of participating American firms indicated that they provided some kind of high school training to their employees (Leontief & Duchin, 1986, p. 93). Given that the current formal educational system has been experiencing increasing dropout rates and declining daily attendance and performance, it is perhaps no wonder that American corporations are increasingly faced with the unpleasant prospect of picking up the slack through recurrent training.

Given cultural differences, training for younger employees cannot help but vary in different countries. The content of European company training for younger employees presents a vivid contrast to that of American companies, and focuses on providing apprenticeships in crafts or trades as diverse as tool-and-die making and office employment rather than on filling the gaps left by formal education. Most of these apprenticeships involve a formal contract between the employee, the organization, and a government agency that outlines the responsibilities of all parties involved. To illustrate, an office worker in Switzerland may begin a three-year apprenticeship after an average nine years of schooling. During the week, he or she will work three to four days and spend one to two days in a trade school learning both general subjects and workrelated topics such as accounting, foreign languages, computer application skills, and employment and trade laws. At the conclusion of the apprenticeship, he or she will then be examined on theoretical and technical aspects of the trade.

Cultural differences, however, are not the only things that affect the content of training. Varying educational standards mean that training needs of the work force in one country are not necessarily the same as those in another, a matter of some concern to organizations operating internationally. When Matsushita was trying to introduce quality control in its U.S. television manufacturing plant, for example, it first attempted to hire high school graduates to fill quality control officer positions — quite a logical move, had the plant been in Japan. American high school graduates, however, were incapable of doing or understanding the necessary statistical work to make quality control a success. The company then turned to college graduates, yet even they needed additional training to match the performance of their Japanese colleagues who only held high school diplomas. The training needs of Americans were, in this case, much greater than those of their Japanese counterparts; for Germans, these needs might be different again. In West Germany, quality control is usually the responsibility of highly skilled tradespeople (*Facharbeiter*) who have the kind of formal schooling-apprenticeship background described earlier. Accordingly, these individuals are likely to need less extensive training *after* being apprenticed to a firm than either Americans or Canadians (see also Chapter 10 for a more detailed discussion of cross-cultural differences).

Nonmanagerial employee training. In most cases, nonmanagerial employee training is limited to the initial training the individual receives when he or she joins the organization. Recurrent training is almost nonexistent and is provided only when new skills are needed because of technology-induced change or because the employee has been given a new assignment. If a company were to decide to link all its computers with a local area network (LAN), for example, its employees would require training to make effective use of the system's capabilities. Unfortunately, research has shown that training will usually take the form of some basic instruction and a manual — not a particularly satisfying method for the end-user, as we have already noted.

Large organizations that can afford to have their own training staff are most likely to offer comprehensive continuous training. The training opportunities these firms offer may be limited to after working hours and tend to concentrate on those subject areas the employer believes are particularly important. In European countries, for instance, employers might (due to the geographic proximity of most European states) focus on courses in foreign languages. The employee must usually initiate continuous education beyond that required by technological change. When there is no fully comprehensive career development system in place, he or she will also decide upon the nature of that education. It is safe to assume that highly motivated employees will acquire training on their own initiative (see Burke & Day, 1986; Gattiker & Larwood, 1986b). If there is a career development system in place, however, company-initiated employee training that reflects the career development plans laid out for the individual can stimulate personal skill building and improved job performance (measured by appraisal data). Such a system will best facilitate the continuous upgrading of employee skills and make for a competent, competitive work force. Most important, the system allows the employee to satisfy his or her growth needs as formulated in the career plan.

Supervisory and managerial training. Most large organizations around the world, such as Arthur Andersen, IBM, Bank of Tokyo, Nestlé, and Dawoe, have their own training seminars for various levels within the corporate hierarchy, through which managers and managerial candidates are prepared for future duties and assignments using practical exercises. Typically these programs are intended to improve both professional and technical skills as well as increase the understanding of, and stimulate interest in, managerial and leadership problems. The desired goal of this process is improved performance and greater personality development (Union Bank of Switzerland, n.d.).

For smaller- and mid-sized firms, however, company-owned training facilities are not very practical, and these organizations may opt to use a solution similar to that used by some business schools, that is, contracting private companies to provide skills training for their supervisors. Again, an organization like DDI might provide this kind of training, using either generic programs or programs tailor-made to the specific needs of the firm. Private contractors may even be used by larger organizations that wish to do their own training. These contractors may develop firm-specific programs and then train the firm's staff to administer them in return for fees or royalties, depending upon the service.

Training and New Technology

The previous sections illustrate that training is an important activity for human resource managers to consider when trying to facilitate the continuous skills upgrading and necessary refreshing needed to retain a competent and skillful work force. One decision the company must make is whether it will support the employee in acquiring company-specific and/or more general skills. Attending external training programs results in a larger general skill component, thereby facilitating the transfer of skills to another employer. The commonly held human capital assumption is that profit-maximizing firms cannot invest in their employees' general training. Human capital theorists (e.g., Becker, 1964) have always assumed that costs of general training, which provides workers with skills that are useful in more than one job or firm, should be borne by employees. The assumption is that productivity gains from this type of training will be equivalent in all firms, resulting in a rising market wage and making it impossible for the firm that has paid for this training to recoup its investment. Becker argues, therefore, that if firms appear to make general training investments, in reality, they are remunerating employees below marginal product. It follows that nontraining firms can pay the prevailing market wage and pirate workers from training firms.

In a high ILMD situation, general training as an insurance mechanism has particularly attractive qualities—it may reduce the voluntary turnover of individuals whose firms financed their education in comparison with individuals forced to pay for their own training (Glick & Feuer, 1984). Feuer, Glick, and Desai (1987) reported that mobility for scientists and engineers is negatively correlated with firm subsidies for postgraduate education. In other words, support reduces voluntary turnover. Thus general training can be of benefit to the firm. Consideration of these results suggests an amendment to the human capital theory.

Pfeffer and Cohen (1984) have also argued that ILMs are, in part, found in firms to the extent that firm-sponsored training is required. Firm-sponsored training (general and firm-specific) may also facilitate intrafirm mobility, especially between divisions. Rivalries between divisions and poaching of skilled manpower, of course, are another set of issues. Nevertheless, training systems in combination with career assessment should help the firm and the employee to prepare for technology-induced change in the workplace, thereby facilitating effective use of the technology and maintaining gainful employment opportunities for the individuals.

Organizational Type and Career Development and Training Systems

Table 8.1 indicates the kinds of career development and training systems characteristic of the four quadrants presented earlier (see Figure 3.1).

Career Development

Firms in Quadrant A usually offer career development opportunities that are linked to ILM rules and regulations concerning seniority rights and job

TABLE 8.1: Four Human Resource Strategy and Internal Labour Market Types: Career Development and Training Systems

Variable	Quadrant A High ILM Determinism, Low Choice in Strategic Human Resource Management	Quadrant B High ILM Determinism, High Choice in Strategic Human Resource Management	Quadrant C Low ILM Determinism, High Choice in Strategic Human Resource Management	Quadrant D Low ILM Determinism, Low Choice in Strategic Human Resource Management
Career development	Job security, limited promotion opportunities; linked to job classification and seniority rights; formalized but limited in scope	Job security, promotion opportunities, and development programs; linked to job classification, seniority rights, and performance; formalized and applied career planning	Individualized development programs; performance-based promotion opportunities and performance; formalized and applied regular career planning	Limited job security and promotional opportunities; informal process, employee-initiated or out of dire organizational need
Training	Due to environmental and union demands	Based on organization of work and nature of task	Based on career inventory, anticipated changes in organization of work, and nature of task	Depends on case and situation, program rarely formalized, support for employee limited, continuous
Type of recurrent education	Job classification-related skills	General skills, breadth, personal career development; used as reward (seniority related)	General skills, breadth, personal career development; often tied to seniority	Usually employee-initiated
Focus	Job group, union members	Group and individual	Individual and group	Group

147

classification. The earlier example of Canada Post typifies this quadrant. Career development for letter carriers or mail sorters is limited, although seniority gives them the right to bid for desirable inside counter jobs. In Quadrant B, a firm with more flexibility in its SHRM (e.g., Siemens, the German electronics multinational) usually offers structured and systematized career development that is affected by ILMD but still allows the high performer to develop independent of ILM rules and regulations if necessary. In this quadrant, the firm uses regular annual career planning, and ILMD does not affect its choice in methods or the flexibility needed to adjust to technological change.

Quadrant C is exemplified by Credit Suisse, which, as we have seen, is currently in the final stages of implementing a career development system that will allow the firm to systematically develop personnel and perform human resource planning, basing advancement on performance rather than seniority and focusing upon the individual rather than on a group or job classification. An organization in Quadrant D, such as the Swiss PTT, may also have great potential for instituting career development but usually will not take advantage of it. These companies do not offer their employees a comprehensive, individualized career planning system, and career development efforts are most often initiated by the employee unless a change in technologies compels the firm to train the work force to understand and use the newly introduced technology.

Training

In Quadrant A, training is mostly based upon union demands and, less frequently, upon the demands of the environment. Recurrent training is likely to be offered, therefore, according to job classification and to be focused on the job group and/or union members. If technological innovation in, say, mail sorting requires additional training for Canada Post mail sorters, it will be provided to all employees classified as mail sorters rather than to any interested parties or those recognized as especially competent. The firm in Quadrant B tries to focus the recurrent training upon the organization of work and the nature of the tasks an individual is performing; it is more willing to provide general skills and is more concerned with using training as a tool for personal development, often providing it as a reward.

Quadrant C firms may be the most innovative and the most likely to implement a training program that also offers continuous upgrading of skills for employees in midcareer. Although Credit Suisse belongs in this quadrant, its training system for midcareer support personnel is very limited, in contrast with IBM or Dow Chemical (also in this quadrant), who have both made substantial changes in this area recently, responding to a need to assure that their work forces have the skills and talents needed to remain competitive. The primary focus of training in these firms is on the individual, with only secondary focus on the group.

Training in firms in Quadrant D is rarely employer-initiated and materializes through employee initiatives or out of a dire need to keep up with technological change. Support for continuous education requested by an interested employee is, therefore, not necessarily demonstrated by a firm's reimbursement of training/education costs.

Career Development and Training Systems: Facilitating Technology-Induced Organizational Change

The previous discussion illustrates how important career development and training systems can be when trying to promote technology-induced organizational change. An important issue, however, that has not been addressed in this chapter to any great extent is how a small- or medium-sized firm may be affected by its quadrant position and ILMD. These kinds of organizations typically do not have a highly developed ILM (Pfeffer & Cohen, 1984) and are, therefore, most likely to be located in Quadrants C or D (see Figure 5.1, and also Table 8.1 in this chapter). Small companies are, however, faced with some unique problems when trying to facilitate technology-induced organizational change.

One of these problems stems from the lack of a formalized socialization program in small and medium-sized firms. Socialization tasks are generally performed by the newcomer's supervisor and peers. Company size also limits career development approaches because there are a limited number of jobs available within a small firm, and existing career move opportunities (lateral and vertical) often depend upon one current job holder's likeliness to leave his or her position.

Table 8.1 suggests that even small- and medium-sized firms must train their employees to help them make the most effective use of new technology, but their size dictates certain financial considerations that must be taken into account when designing training programs. Based upon recruiting and selection efforts, current employees may already possess certain skills, and an extensive performance appraisal system can provide the organization with the necessary information about what type of technology-related training the work force is most likely to need. If technology-related training is given, it is very likely to be provided either on the job or by vendors. The critical point that arises from this discussion is that all the human resource activities outlined so far — employment, recruiting and selection strategies, socialization and appraisal systems, training and career development systems — are interrelated and need to be consistent to enable the firm to manage its ILM, maintain or expand its choice in SHRM, and initiate or respond to technology-induced organizational change effectively.

Summary

Internal labour market determinism and strategic human resource management are important factors in determining how the organization can implement career development and training systems. While high ILMD tends to prevent the organization from offering extensive career development or recurrent training systems, greater freedom in SHRM can foster them. In whatever quadrant an organization falls into, however, new innovations are constantly redefining the way organizations run. Because employees at every level can adapt better if they are well versed in the technologies with which they are expected to work, most firms should increase their training and continuous education efforts for all personnel, regardless of their ranking in the organization. Postsecondary institutions should also respond to these training needs because they are often looked to as providers of this kind of training.

The rapid introduction of new technologies, and continuous innovation in this area, make career development and training systems an important component of any effective human resource strategy in an organization. ILMD and environmental determinism demand a long-term outlook when it comes to career development and training. Such a farsighted approach facilitates technology-induced change and enables employees to adjust without experiencing negative outcomes, such as reduced skill requirements, which can lead to boredom, or job losses, which invoke a pessimistic attitude towards technological change (see Part III of this book; also Rogers & Friedman, 1980). We have already focused on employment strategies, recruiting and selection systems (Chapter 6), and socialization and performance appraisal systems (Chapter 7), which, as illustrated, are influenced by choice in SHRM and the ILMD experienced by the firm. As discussed above, career development can help the organization face technological change and adapt by facilitating training efforts to use new technology. It is especially important to note the way all of these systems work together to enhance organizational effectiveness.

The next chapter expands this system of interrelationships by illustrating how compensation systems are affected by ILMD and choice in SHRM. A combination of the insights associated with each of the perspectives in the previous chapters is used in an attempt to account for the emergence of organizations that effectively adapt to technological innovation by using compensation system components that fit their ILMD and choice in SHRM.

9. Compensation Systems

Compensation systems can make the difference between desirable and undesirable organizational cultures and often determine the attractiveness of the organization as a work environment. The *remuneration system* is part of the overall compensation package offered by the firm, and it provides the salary or wage that is usually the major portion of the compensation the individual receives from the firm in return for his or her services. The *reward system*, a subsidiary compensation system, provides incentives that direct and motivate employee behaviour towards the end of improving individual and collaborative performance (Kerr, 1975).

In our initial look at the effects of internal labour market determinism and strategic human resource management on compensation systems in Chapter 5, we discussed compensation in terms of an organization's reasons for providing it but not the specific type of compensation and rewards. Instead, we outlined how ILMD and choice in SHRM as overall parameters might affect compensation and, especially, how organizational culture may be reinforced by using certain types of reward and compensation strategies (see Tables 5.1 and 6.1).

The law requires an organization to provide the salary/wage and fringe benefits that constitute *remuneration* but it is not legally obligated to provide *rewards*, which are given as incentives for better performance. Although we should make an effort to continue thinking of compensation as fitting in either of these two classes, in this chapter we will learn to categorize compensation according to its *nature* and examine how organizations might reward their employees with *pay, job content*, or *status differentiation*. Any one of these types of compensation might be used by either a remuneration or a reward system (one kind of job content compensation might, for example, be a legal obligation of the employer to the employee and another might be used to provide incentives or reward performance). As it happens, however, pay compensation is probably more likely to be used as remuneration than status differentiation, which is in turn more likely to be used as a reward, and so on. The flexible nature of the relationships between remuneration and reward systems and pay, job content, and status differentiation compensation should be clearly understood at this point, so that there is no contradiction when we discuss any one of these types without distinguishing between *remuneration* and *rewards*. Experiencing a given level of internal labour market determinism and choice in strategic human resource management, the average firm usually chooses to use certain methods of making pay, job content, and status differentiation compensation to its employees.

From the organizational point of view, compensation systems, like the organizational systems discussed in the previous chapters, should be consistent with long-term strategies while reinforcing appropriate work behaviours. The ILM is less concerned about such consistency, however, and more concerned with ensuring that these systems respect its rules regarding demarcation lines, seniority, reentry, and so forth. If there are strong union ties, it will be more successful in doing this, because unions excel in controlling compensation and commonly negotiate contracts for their members according to industrywide agreements that the organization would find difficult to dispute. The organization that *can* defy union influence and insist on an individually negotiated contract (as we saw Burlington Northern attempting to do in an earlier example; see Chapter 5) will gain a greater flexibility in setting compensation strategies akin to that experienced by organizations in quadrants that are not unionized.

The effects of ILMD and SHRM on compensation systems can be illustrated by looking at the example of the Swedish Metal Workers' Union (SIF), which called a three-week strike at the beginning of 1988. One of the major concerns of negotiation was the union's desire to obtain the right to directly influence organizational compensation strategies and policies. The compromise approved by employers and SIF gave the union less opportunity and legal justification to influence compensation policies than it had hoped. Had the striking workers received the right to directly affect compensation strategies, the resultant high level of ILMD and decreased choice in SHRM would have meant that salary increases and bonuses would probably not be as closely tied to performance appraisal data, profitability, and career development considerations as they had been.

Types of Compensation Systems

Pay Compensation

Pay is a somewhat ambiguous term as it may take the form of wages, fringe benefits, cash bonuses, and so forth (Heneman & Schwab, 1979). *Pay levels* vary both between and within organizations, and *pay structures* are defined by the relationship of these levels across occupations, jobs, grades, and other classes within the firm. The set of contingencies that determines pay levels and changes in these structures is called the *pay system* (Heneman & Schwab, 1979). From the organizational viewpoint, the most rational of these contingencies is *performance*. Because subjective interpretations of performance might differ, some pay systems establish objective definitions of performance with contingent pay levels. Production employee incentive plans, sales commission plans, and some executive compensation plans are features of such systems. These systems quite naturally operate most successfully if the employee understands the definitions of performance the organization uses (which is to say, if the employee understands what the organization expects from him or her) and the contingency between performance and compensation. However, as Chapter 7 has shown, organizations often fail to clearly formalize and structure their performance appraisal systems, with the result that the employee may not see the relationship between the bonus or salary increment received and his or her own performance. Furthermore, for top managers, subjective elements in a performance appraisal may be quite important when trying to use a merit pay system (Pearce, Stevenson, & Perry 1985), thereby sometimes making it more difficult for the assessed to see the link between performance and rewards.

For occupational groups whose performance output is not easily measured, or is heavily influenced by the actions of other occupations or jobs (as is the case with many managerial positions), it may be impossible to establish an objective definition of performance. For these groups, organizations may establish *merit pay systems*. Generally speaking, these systems differentially reward employees on the basis of relative and (occasionally) absolute assessments of their performance, most do not specifically define performance criteria, and most allow managers to consider nonperformance factors in determining the magnitude of salary increases. The effectiveness of merit systems is uncertain and has been little researched (Dyer & Schwab, 1982). In one of the few studies that have been done, Pearce, Stevenson, and Perry (1985) found that, in five federal government agencies in the United States, the implementation of merit pay had *no* significant effects on organizational performance. This may be because merit pay systems are usually intended to increase *managerial* performance, which accounts for only 10% of the variance in organizational performance (Pfeffer & Salancik, 1978). Managers who realize they cannot increase their rewards no matter how hard they work will see no carrot but only an intolerably long stick (Pearce, Stevenson, & Perry, 1985). Still, the merit pay system is useful insofar as it encourages members within the organization to cooperate with each other rather than compete, and the gradual increase in performance on which the resulting favourable organizational culture would be contingent might, in the long run, do the organization more good than an increase in organizational performance dependent upon the ambition of one employee who will probably not be with the organization through its entire existence.

In addition to performance, managers and compensation professionals usually consider other factors when determining pay levels. Some of these are impersonal and are not determined by particular jobs or job holders, such as the size of the salary increase budget and cost-of-living increases; others are based on resource dependence or upon the individual characteristics of the job holder or job, such as organizational tenure, the demand in the external labour market for a given job, and the disruption that would be caused in the organization by the job holder quitting (Fossum & Fitch, 1985). This disruption will depend on how critical the position is to the organization (*criticality*), which is determined by its importance to the firm, the ability it gives its incumbent to cope successfully with the uncertainty and the pervasiveness of the activities it involves, and the availability to the organization of alternative means of coping with this uncertainty. Discrimination, unfortunately, might also be a contingent for determining pay. Jobs held mostly by women, for example, tend to have lower salaries than those held mostly by men (Kemp & Beck, 1986). Although most research about discrimination and its effect upon compensation has been done in North America, numerous studies report data indicating that women face pay discrimination in such countries as West Germany (Krebsbach-Gnath et al., 1983), Finland (Kauppinen-Toropainen, Kandolin, & Haavio-Mannila, 1988), Japan (Edwards, 1988), and Switzerland (Kugler, 1988).

No matter how carefully the organization attempts to determine fair and just compensation (taking into account all of the above factors), it cannot ensure that the employee will receive the increases that he or she believes are deserved. Pay and reward satisfaction can be lowered even for individuals in critical positions (who generally receive a higher remuneration package than those in jobs of similar content in less critical positions for the firm) within the organization if ample alternative employment opportunities are available or if they possess skills that are in high demand in the external labour market (see Arnold 1985; Windolf, 1986). Hence, the individual may start to compare his or her remuneration and reward package with the one received by identical job holders in other firms or relate it to the remuneration package offered to the individual by a competitor for changing employers. In both cases, the remuneration and reward package may be substantially higher than the current package, thus possibly reducing current pay and reward satisfaction.

The contingents determining pay levels discussed so far tend to affect wage levels in particular, yet the costs of labour alone give little indication of the total pay expenses the organization usually faces. Additional costs like medical and unemployment insurance and pension and dental plans (usually percentages that increase with the hourly wage and are fixed by some sort of regulation) add a great deal to the pay package. In Japan, at the lower end of the scale, these costs typically represent 28% of the hourly wage, while in Canada and the United States, the percentages are 36% and 40%, respectively; at the higher end of the scale are West Germany with 81%, Italy with 94%, and Austria with 95%, to mention only a few examples (UBS, 1985). As the situations of the countries at the high end of the scale might suggest, additional costs can be a matter of even greater concern to the organization than the base labour costs themselves. The expense and restrictions of regulated additional costs can limit the organization's options as far as offering higher hourly wages or less orthodox fringe benefits. Neither are these regulated costs wholly beneficial to the employee, because the security they offer simultaneously limits the employer's choices in compensation types.

Additional wage costs have also led to an increase in employment strategies that are not to the worker's benefit, such as nonguaranteed part-time work on term contracts (e.g., the Dutch example discussed in Chapter 6). Moreover, high quasi-fixed wage costs increase the ratio of part-time versus full-time employees (Montgomery, 1988). If the ILMD level permits the increase of such types of employment contracts, additional pay compensation costs are reduced dramatically—the employer must pay on those only benefits required by law (e.g., unemployment insurance). Moreover, flexibility in structuring pay compensation may be increased by such an employment approach because these jobs are not likely to be part of the ILM contract agreement.

Profit-sharing programs might represent one strategy for reducing these additional costs. For firms, a profit-sharing arrangement reduces the fixed hourly wage and thus the fringe benefit costs; in addition, it involves cyclical flexibility of labour costs (higher profitability results in higher labour costs equals profit payout) and, therefore, greater stability of profit levels and rates (Nuti, 1987). For the worker, the greatest benefit of this increased stability of profit levels is that it increases job security. However, employees must accept greater earnings dispersion and incomes that are more likely to fluctuate, along with the decrease in the benefits profit-sharing is designed to reduce. Though research has shown that profit-sharing programs boost productivity, unions are often against such programs because they tend to increase cooperation between employees and management (Nuti, 1987).

ILMD and pay levels. A relatively strong ILM may play a large role in determining pay levels, and its concern with fairness will cause it to fight against pay discrepancies caused by the organization's use of performance, resource dependence, discrimination, and gender to determine pay levels. To this end, it will demand that the organization pay according to the doctrine of *equal* or *comparable worth*, which is to say it will demand that employees holding different jobs judged to be of similar value will be paid the same regardless of their sex and no matter what the external market value of their skills. The ILM will also usually determine the relative worth of employees using job classifications based on skills and education.

In an organization that is forced by its ILM to use this somewhat awkward scheme to determine pay levels, the earlier mentioned resource-dependence model proposed by Pfeffer and Davis-Blake (1987; see Chapter 7) may be untenable because the firm will be unable to pay according to the importance it places on a position or according to the demands of external labour markets for an employee's skills. Any attempt to do so is likely to result in grievances or lawsuits (see Cooper & Barrett, 1984). This is unfortunate, because the company that can pay according to resource dependence is more likely to attract and retain personnel with the skills necessary to work with new technologies and will be more capable of offering better opportunities than the outside market (which should also increase the employee's behavioural commitment to the firm; O'Reilly & Caldwell, 1981). In addition, resource dependence provides psychological benefits for the employee that the organization can exploit. The individual who is conscious of the fact that he or she is being paid more because his or her job is important is more likely to feel committed to the firm, and other employees, knowing that the possession of vital skills places them higher in the pecking order, are encouraged to acquire such skills so that they too can enjoy the sensation of being "needed" (see Pearce, Stevenson, & Perry, 1985; see also Chapter 8). An inability to pay according to resource dependence, on the other hand, makes the adaptation of innovation far more difficult. The company is less able to acquire the skills to work with innovations because they often replace rather than supplement existing skills, which means that the job classification of the individual possessing these skills will not change. As we have seen, if there is no change in job classification, the comparable worth doctrine rules that there can be no change in pay, even though the market rate for skilled workers may be much higher than is standard within the organization.

One way the organization might be able to overcome this handicap is through the use of freelance or part-time employees. Because they are usually not connected with the ILM, these employees, if they possess skills that are relatively scarce in the labour market that are needed by the organization, can be paid the market wage for their services, even if it is higher than that paid to other part-timers or even full-timers. In this way, the organization can acquire the skills it needs while still technically abiding by the ILM's equal worth policy. An organization with no such alternatives, however, might be forced to provide currently employed individuals with vital new skills through in-house training (Shaiken, Herzenberg & Kuhn, 1986). Unfortunately, although there are advantages to this type of training, as we saw in Chapter 8 — namely, that it can be designed to be so company-specific that the transfer of skills learned to other firms would be difficult, reducing the likelihood of turnover and thereby discouraging a demand for higher wages to go along with new skills — in-house training consumes valuable time and requires the presence within the organization of someone already equipped to provide it. For an organization that is relatively small, this may be altogether unfeasible.

Pay structures. One of the biggest drawbacks for the organization forced to use a comparable worth philosophy is that such a doctrine also forces the organization to pay higher wages overall. This has led to the creation of *two-tier wage structures* that allow the organization to concede to union demands for comparable compensation while still paying differentiated wages. In these increasingly common structures, the top level of pay for new employees is substantially lower than that for long-time employees. Thus personnel with the same job titles and duties receive different pay for similar outputs of effort. Most such structures, as Jacoby and Mitchell (1986) found, are temporary arrangements in which newly hired employees can eventually work their way up to the higher wage scale of veteran employees. Only 30% to 40% of these structures were permanent. Generally speaking, two-tier structures do not affect part-time and term position holders because these employees are typically not an official part of the ILM structure, which gives the organization a freer hand in determining their levels of pay.

Two-tier structures violate the basic union tenet of equal pay for equal work and are, therefore, likely to affect employees' perceptions of equity and justice (Gattiker, 1987b). Unions fear that these structures may create more privileged and less privileged classes of employees and cause the work force to split along these lines, weakening labour's bargaining positions. Still, the lower wage costs of these structures to the firm can increase its competitiveness and make its adaptation of innovation easier, which tends to make jobs more secure (e.g., Nuti, 1987; UBS, 1985). Because job security is one of the main goals of unions, they must be willing, however reluctantly, to concede to the existence of these structures.

Pay structures are not always based on as simple a concept as tenure, and quite often the pay structure of an organization is primarily defined by pay levels that differ according to hierarchic positions. Pay structures vary between organizations because there will always be various degrees of hierarchic pay differentiation in different organizational settings. Stinchcombe (1963) has argued that organizational wage inequality is greater when individual activities are talent complementary rather than talent additive. When activities are *talent complementary*, an individual's performance can add disproportionately to organizational output; when activities are *talent additive*, each person's output contributes more or less equally to total organizational output, and disproportionate contributions are not possible. Accordingly, Abrahamson (1973) theorized that, because research universities have higher levels of talent complementary activities than teaching institutions, they should have more wage dispersion across hierarchic levels. He tested this by examining the wage ratios between full and associate professors, full and assistant professors, and associate and assistant professors. The findings confirmed that wage inequality was greater in research universities than in teaching institutions, and they thereby agreed with Stinchcombe's theory. However, it is important to realize that, if the organization does not consider an employee's output to be disproportionate, his or her pay will not differ as greatly from that of his or her coworkers as outsiders who perceive a disproportionate output might expect. A university that for some reason does not value research might not pay the market rate for a top researcher on staff. The average importance of a job in the general market is not, therefore, the final determinant of how important a job is to a specific organization. In support of this, Pfeffer and Davis-Blake (1987) found that the incumbents of positions that were more important in private universities (e.g., fund raising officer) than in public institutions were paid comparatively more in private than in public settings.

Through their control of the ILM, unions try to influence pay structures in such a way that the differences between pay levels will be reduced. Collectively bargained standard wage rates and seniority-based rates reduce the dispersion of wages within the organized sector and within establishments, and unionism reduces inequality in earnings by more than offsetting the dispersion of wages across industries. Quan (1984) showed that wage earnings in ILMs, and for unionized employees in particular, were more equally distributed among recipients than nonunion earnings.

Job Content Compensation

When the average person considers the topic of work compensation, he or she is perhaps not likely to consider the possibility that such compensation might involve the improvement of working conditions as well as the provision of financial benefits. *Job content compensation*, however, may in the long run prove to be more important to the organization's human resource management strategy than pay compensation, and so it must be an important part of any serious discussion concerning compensation systems.

According to the social information processing model, perceptions of and approaches towards tasks are at least partially a function of social cues in workplaces; this implies that perceptions of job content are based upon the individual's interpretation of events or situations within his or her work context. In the case of an employee in a large organization, the work context is the corporate structure (see Fishbein, 1967). Pfeffer (1981b, p. 10) provides perhaps the best summary of the social information processing model:

> First, the individual's social environment may provide cues as to which dimensions might be used to characterize the work environment.... Second, the social environment may provide information concerning how the individual should weight the various dimensions—whether autonomy is more or less important than variety of skill, whether pay is more or less important than social usefulness or worth. Third, the social context provides cues concerning how others have come to evaluate the work environment on each of the selected dimensions.... And fourth, it is possible that the social context provides direct evaluation of the work setting along positive or negative dimensions, leaving it to the individual to construct a rationale to make sense of the generally shared affect reaction.

Griffin (1983) tested the social information processing model in a field experiment by training first-line supervisors to provide positive social cues to their subordinates about their jobs, comparing the effects of those cues on task perceptions and attitudes with the effects of objective task changes, and analyzing the interactive effects of the cues and changes. The results indicated that the social cues were just as powerful as the objective changes in altering perceptions and attitudes.

The success of an organization in determining the kinds of incentive it needs to improve employee performance will, according to choice-based theories of work motivation, depend upon an accurate assessment of employee preference for different job attributes (Lawler, 1981). Shapira (1987) used the trade-offs method among several characteristics associated with managers' current job positions. Rather than asking executives how much challenge they would like in their jobs, he asked them how much they would be willing to give up in pay (or pay raises) in return for more challenge. Other characteristics were (a) influence on company's policy, (b) status, and (c) authority. Salary increase was valued less by most of the executives than obtaining a more interesting and challenging job or having more influence on the company's policy.

Jobs can usually be discussed in terms of a set of predetermined characteristics such as skill variety, task identity, task significance, autonomy, and feedback. Shapira's study suggests that a positive interpretation of these characteristics by a manager may offset the effects of a possible lower remuneration level, and Gattiker and Larwood (1986a) reported that perceptions of job characteristics were important in explaining why support personnel, who often perceive their jobs as not offering enough desirable characteristics, may feel less successful in their careers than managers.

Griffin, Bateman, Wayne, and Head (1987) examined the combined effects of social information processing and job characteristics on task perceptions and responses. Their data supported a general, integrated perspective on perceptions of tasks and responses to them, which is to say that objective facets of workplaces and social information jointly determine perceptions and effect. For an organization this means that providing a suitable reward for work well done is often not enough in itself to keep the employee happy, because his or her opinion of the reward may be altered by talking to peers who may perceive its value differently (see also Chapters 10 and 11).

The interpretations discussed above may influence employee attitudes towards technological change, and because social information processing differs from firm to firm, different attitudes will result. For example, Rodrigues (1988) reported that Brazilian supermarket employees felt that new information technology decreased control by others and gave them more autonomy, although it also increased work volume and rhythm. On the other hand, Brazilian bank employees reported that they were more satisfied yet also felt more stress and that technological change made work more abstract.

Rodrigues suggests that introducing technology may affect outsiders' interpretations of job content differently from insiders' and that what may be seen as positive change in one organization or industry will not necessarily be seen as such in another. This is especially likely to be the case if the relative influence of ILMD and unions differs between two organizations. Because the subjects of the Brazilian study differed despite the fact that the influence of ILMD and unions upon the organizational adaptation to new information technology was for all practical purposes nonexistent, it is safe to assume that when this influence is great, it will affect information processing to an even greater extent and that, if the information provided by the union is negative, its processing by the individual may in turn lead to negative interpretations of job changes.

Kirmeyer and Shirom (1986) investigated perceived job autonomy in unionized organizations and found that unionization and employees' freedom in deciding how to do their work were indeed negatively correlated. It is likely that this will usually be true and that job autonomy (i.e., both perceived and actual) will be lower for employees who work in a firm with a powerful ILM than it is for employees working in a firm without one. The presence of an ILM might diminish job autonomy because management may resort to formalization of policies and practices (Gattiker, 1988a) to ensure the uniform treatment of the rank and file across jobs. The organization can avoid overly rigid ILM rules and regulations by emphasizing employees' involvement in decision making (Windolf, 1986) and by using broad-banded job classifications (Stark, 1986).

Although ILMD and unionization in particular limit employee autonomy, union workers will not necessarily *perceive* that they possess less autonomy and control than employees in firms unaffected by ILMD. Information provided by the union can affect the employee's information processing and cause him or her to interpret seemingly negative job characteristics as positive. It may also happen, however, that the union raises expectations on which it cannot deliver, thereby affecting morale negatively and causing employees to perceive changes negatively, which is what happened in the case of the SIF strike in 1988.

Job compensation is likely to be affected by social information processing, perceptions, and attitudes held by employees (see also Part III for an extensive discussion of this issue). Nonetheless, addressing individual needs and perceptions is easier in an organization that does not have specific job classification systems to which it must strictly adhere. High ILMD limits individualization when it comes to job content compensation; more flexibility would allow the manager to respond more effectively to such things as employee career development needs.

Status Differentiation Compensation

Surveying several firms, Kerr (1985) found that, for managers, three types of compensation systems based on organizational characteristics appear to exist. In a hierarchic bonus system, rewards may comprise up to 20% to 30% of pay, and salary increases tend to be based on performance and tenure. In such a system, stock rewards may also be used, and these will usually be even more hierarchically keyed than salaries. A second kind of system is performance based and stresses precise definitions and measurements for performance. Bonuses tend to range from amounts equal to 40% of the employee's salary to amounts that are unlimited. A third system represents a compromise and falls between the poles of the other two. The range of bonuses is wide — from 20% to 70% of salary. Raises are based primarily on performance, but rank and tenure are factored into salary adjustments. Companies in this group tend to stress promotion for developmental purposes rather than for vacancy filling.

The reason we are interested, at this point, in the three types of compensation systems Kerr describes is because the hierarchic bonus system had *status differentials* present, differentials that were deemphasized by both the performance-based reward system and the compromise system. This seems to indicate that *status differentiation* is more important in some organizations than in others — most likely in organizations that do not experience high levels of ILMD. Because performance-dependent financial bonuses are likely to be uncommon in organizations with powerful ILMs, the organization that wishes to differentially reward higher achieving employees must often resort to using status compensation. Status compensation can give the organization a loophole for avoiding ILM restrictions governing compensation by allowing the firm to compensate the employee for performance in a way that is unaffected by the official compensation package and that may still be justifiable as necessary to the worker's job in some way. Status indicators such as office space, furniture, and travel arrangements might, therefore, be used for compensation purposes, because such things can be justified as being of benefit to the company as well as to the employee — quality furniture in a favoured employee's office, for example, perhaps being explained by the fact that customers visit this office more frequently than others.

Though status differentiation is usually more important in organizations with high ILMD, organizations with more freedom in designing compensation packages may also use this kind of compensation. Offering status differentiation as well as pay compensation enables a company to offer a more attractive package — the prize at the bottom of the Cracker Jack box, so to speak. For example, in a discount brokerage such as Schwab Securities, a U.S. firm, the position of broker is probably the most critical to the firm's success (see Pfeffer & Davis-Blake, 1987). The company's resource dependence on the broker and the external demand for his or her skills — especially if he or she has a customer base made up of large institutional and private investors and is a particularly successful salesperson — make it important that he or she is more than adequately compensated for these skills. Rewards for brokers must, therefore, be clearly performance based. Part of Schwab's solution to its reward problem was to offer employees foreign-made cars, the value of which symbolized the broker's status in the firm.

In an organization like Schwab, the status differentiation system can become institutionalized because the firm, free from a high level of ILMD, does not have to be covert about performance-based compensation systems. For this reason, Schwab's status differentiation system was drastically affected by its takeover by Bank of America, an organization with a high level of ILMD. Bank of America could not justify a formalized status-reward system because any such system would have to be incorporated into the formal compensation package. This company also could not justify rewarding employees according to resource dependence rather than job classification. Because Bank of America only provided two types of company cars (both of which were American built and far less expensive), and only gave these to employees whose jobs gave them some pretext of needing them, the expensive cars driven by Schwab's brokers (some of whom drove better cars than their new superiors in the parent company) were an embarrassment the ILM could not accept. The result was the termination of Schwab's status differentiation system, which led to Schwab losing some of its brightest and highest-performing employees. Eventually, Bank of America decided to sell the company back to Schwab, for several reasons. One major factor was the cultural clash occurring between the two firms' work forces.

Compensation Systems and New Technology

How do compensation systems relate to technology acquisition and its introduction into the workplace? During the technology introduction phase, the individual may be required to acquire new work skills, and at later stages certain procedures will undergo further change to make the most effective use of the new technology. Employees who gain skills that pertain to the technology should be compensated to reflect their additional knowledge. For instance, Canadian Airlines International has started to offer an information service to its frequent flyers, allowing them to inquire about their frequent flyer accounts (listing the miles accumulated on the airline and its partners) and ask for rewards by using a modem to access the firm's computer. It is very likely that, in the near future, frequent flyers will also be able to purchase their tickets via this computer link, which will save the firm the 10% commission it currently pays to travel agents.

Naturally enough, the advent of this new service has led to occasions when the airline's customer representatives must respond to client's queries about how to use the computer hookup. In order to do so effectively, customer representatives' skills must now include sufficient computer knowledge to help frustrated customers obtain the firm's services and product by using the company's computer. How should individuals with such additional skills be compensated?

These and other questions must be addressed within the framework outlined in Chapters 3 and 4. Compensation practice depends upon the ILMD and the choice in SHRM the firm enjoys. These issues are considered further in the section below.

Organizational Type and Compensation Systems

Table 9.1 and the following discussion examine how the four different types of firms distinguished in Figure 5.1 manage pay, job content, and also status differentiation systems, and how compensation affects organizational adaptation of innovation.

Pay Compensation

As we would expect, in Quadrant A, the influential ILM dictates pay systems and their levels, forms, and structure. Membership and contractual obligations probably are the major factors determining the employee's financial rewards in an organization in this quadrant. Quadrant B firms are more fortunate in that they enjoy more flexibility in implementing compensation systems. Although wage levels continue to be influenced by external forces as well as by internal equity, the company will be able to use two-tier wage structures to offset the effects of these forces, and the company's greater freedom means that it will be able to pay more to employees in positions that innovations have made critical. This is fortunate, because this quadrant is more likely to be washed by the tides of innovation than Quadrant A, and resource dependence is thus more likely be a problem for the company.

The firm in Quadrant C may make use of two-tier wage structures as well as resource-dependence approaches, and the forms of compensation may vary across organizational levels and job classifications. The criticality of individual positions and the demand for an employee's skills in external labour markets are likely to play an important part in determining compen-

TABLE 9.1: Four Human Resource Management Strategy and Internal Labour Market Types: Compensation Packages

Variable	Quadrant A *High ILM Determinism, Low Choice in Strategic Human Resource Management*	Quadrant B *High ILM Determinism, High Choice in Strategic Human Resource Management*	Quadrant C *Low ILM Determinism, High Choice in Strategic Human Resource Management*	Quadrant D *Low ILM Determinism, Low Choice in Strategic Human Resource Management*
Financial	Based on membership, contractual obligations, tenure, and hierarchic position	External and internal equity, nature of work, and ease of replacement; rewards are based mostly on job classification and individual performance Possible two-tier wage structure	Nature of work, ease of replacement, and past performance record; stock options, profit sharing, and bonuses all based on individual performance opportunities Two-tier wage structure likely; resource-dependence approach	Terms of employment
Job content	Limited challenge and responsibility	Challenge, responsibility, feedback, and recognition	Challenge, responsibility, feedback, and recognition, freedom and meaning, enrichment/enlargement work/job functions	Limited challenge and responsibility, freedom and recognition
Status differentiation	Privileges and facilities based on hierarchic level/job classification and tenure	Limited privileges and facilities based on hierarchic level/job classification; titles and committee memberships; privileges for critical positions	Very limited privileges; few titles and committee memberships except for critical positions with job skills that are in high demand	Except for top management, few if any status rewards

Compensation Systems 167

sation packages. In Quadrant D, organizations' compensation packages also vary widely, although the lack of any true compensation strategy usually means that their forms are described only by the terms of employment (i.e., the contract between the employee and the firm).

In the Swiss restaurant chain, Mövenpick, in Quadrant C, financial rewards are usually resource dependent, and profit-sharing for managers is based both on their own performance and on the profitability of the restaurants they manage. The manager might also be rewarded for learning how to handle innovation, with his or her success in doing so affecting the magnitude and type of reward. If the firm were in Quadrant B, internal and external equity considerations as well as job classifications would be more likely to determine (or at least influence) remuneration and rewards, and the firm would be far less flexible in devising compensation for resource dependent positions. In Quadrant A, resource dependence would not even be considered as a factor, because all compensation would depend on contractual obligations. In Quadrant D, the company might pay staff according to individual contracts and would be able to compensate them based on resource dependence if it wished. However, because this quadrant is relatively unaffected by innovation most of the time, it is less likely that a company here would see an urgent need to hire or retain any employee simply because he or she possessed innovative new skills. Based upon this example, it is obvious that technology acquisition and organizational adaptation are greatly affected by the firm's compensation systems.

Job Content Compensation

Earlier we saw that high ILMD results in greater levels of structure for a number of variables responsible for lower levels of self-determination. Job content, therefore, gives an individual more freedom in Quadrant C than in either Quadrants A or B. Although ILMD is low in Quadrant D, a bureaucratic approach may limit this freedom considerably.

To illustrate this, we will again consider Mövenpick. This firm makes a serious effort to give people more responsibility and variety and also gives employees the option of enlarging their jobs vertically (i.e., participating in more of the steps in the production of a product — a process that also provides more potential for decision making). These rewards are usually based on performance appraisals and are linked to the career development plan. By changing job content through enrichment and enlargement, the individual is rewarded while simultaneously being prepared for more challenging future assignments.

In Quadrant B, Mövenpick would be limited by ILMD in changing work/job functions and levels, and job enrichment and enlargement would not seem feasible, although challenges and responsibility would still be perceived and the employee would still receive recognition for his or her work. If Mövenpick were in Quadrant A, job rules and classifications would make any of the above described types of job content rewards impossible, because changing job content would necessitate the reclassification of the job, which would in turn require a salary increase. If the current employee were to be replaced, the job would need to be reclassified yet again in order to start the new employee off with a lower salary. The paperwork all of this would require would be enough to discourage even the most well-meaning employer from providing job content rewards. In Quadrant D, the situation is likely to be very similar, although the firm would be more likely to offer the employee some recognition and a certain latitude in his or her work.

Technology acquisition and subsequent organizational adaptation change the content of jobs considerably in such diverse industries as software engineering (Kraft & Dubnoff, 1986) and manufacturing (Shaiken, Herzenberg & Kuhn, 1986). The company free to change job content without outside interference, like Mövenpick, is better able to facilitate this process.

Status Differentiation Compensation

In Quadrant C, the resource-dependence approach that determines pay remuneration may also create status differentiation for critical positions perceived as being talent complementary (adding disproportionately to total organizational output). Although firms in Quadrant B might try the same, the ILM governance structure might limit their choices somewhat. High ILMD in Quadrant A causes status differentiation to be based on hierarchic levels, job classifications, and organizational tenures. In Quadrant D, status differentiations are few, although they may be present in top management.

In Mövenpick, committee membership (a common status symbol) is usually determined by the criticality of the potential member's input to the success of the committee's work. Status differentiations are minimal. Although there are no company cars offered to employees in this organization, higher-ranking individuals in administration receive private offices as opposed to being forced to share in the communal office space used by others in the company's head office. In Quadrant B, the company would probably have an executive dining room, and status-based perks would be more common in order to compensate for the company's reduced capacity to provide financial remuneration. In Quadrant A, the company would not be able to provide status rewards except on the basis of job classifications or covertly; and if it were in Quadrant D, Mövenpick would offer status rewards to its top management only.

Compensation Systems: Facilitating Technology-Induced Organizational Change

How can compensation systems facilitate technology-induced organizational change? First, they must be consistent with the employment strategy and recruitment and selection systems. If the company has enough choice in the matter, it will try to attract individuals to the firm who possess vital skills and are remunerated based on their criticality to the company's success. Second, socialization and appraisal systems must be designed to facilitate technology-induced change. For instance, appraisal systems must have a technology component allowing the evaluation of the employee in relation to the technology needed to perform his or her tasks. Moreover, the reward system must be linked to the appraisal system, which means that bonuses and salary increases should in part reflect the technology dimension (i.e., effective use of new technology).

Another important issue is that career development and training systems must be designed to allow the individual to realize the opportunities and challenges that new technology provides. Further, compensation must reflect the fulfillment of potentials reached through career planning. Finally, the compensation system is important on its own. We hope that it gives management the opportunity to make effective use of the previously mentioned systems by rewarding individuals who try to adjust themselves (e.g., skills upgrading) and take on the challenge of managing new technology effectively or become a valuable source of advice for customers if problems arise. Thus consistency between the compensation system and other human resource management systems is the key for facilitating technology-induced organizational change.

Summary

In this chapter we have examined the relationships between internal labour market determinism and strategic human resource management and compensation systems. As we have seen, an organization's success in adapting to innovation is often dependent upon its compensation systems. In Quadrant C, the organization's freedom in selecting compensation systems is an important part of the formula that allows it to adapt so much more effectively than organizations in other quadrants. To illustrate, we will draw on the example of the new Airbus 320 adopted by Air France and Swissair (see also Chapter 3, where this example was used to illustrate proactive adaptation). This plane is so high-tech that essentially only two pilots, instead of the usual three, are needed to fly it. Consequently, Swissair asked for a version that took advantage of this fact and was specifically designed for a two-man crew. The airline was able to adopt these planes without much dissent from its pilot association not only because it guaranteed existing jobs but also because it compensated pilots for the change by asking for their input long before the new planes went into actual service. They also rewarded them according to the added responsibility and increased levels of stress that being part of a two-man crew implied. By compensating the pilots both financially and with involvement in the actual process of adopting the planes, the airline was able to introduce this innovation smoothly and increase revenue per passenger mile due to lower crew costs.

In contrast, Air Inter of France in Quadrant A, when adopting the same technology, informed its pilots in 1987 about the change that was to occur in 1988 without making any initial effort to recompense them for these changes, guarantee them job security or involve them in the process of making the change. This led to strikes in 1987 and again in 1988 because the high level of ILMD affecting this organization required a reclassification of the pilots' jobs based on the change in job content. Although the organization was unwilling to do anything to facilitate the change, cockpit crews demanded an additional work space for a third person for whom there was no work. Compensation could have greatly eased French pilots' qualms about the technology and averted the difficulties described.

Incremental innovation, of course, is easier to accomplish, because changes are made more gradually and affect job content less substantially than in the above example. The reward system is also more easily adjusted for the purpose of helping employees master the new situation while increasing their extrinsic and intrinsic motivation to do so (Deci, 1975), although this will still be more difficult in organizations with high levels of ILMD.

The previous chapters in Part II of this book discussed how a company's approaches to selection and assessment, socialization and performance appraisal, and career development and training systems function depending upon the experienced choice in SHRM and ILMD by the firm. Moreover, these systems influence the firm's compensation systems. It is clear that all of these systems tend to support each other — how, for example, candidates

Compensation Systems

for positions are selected and trained in such a way that they come to accept the intrinsic logic of the remuneration system employed by their organization and how this remuneration system in turn contributes to the socialization of the individual in the attitudes characteristic of his or her peers within the company.

Part II of this book, therefore, suggests that, in a systematic analysis, the socialization and performance appraisal, recruitment and selection, career development and training, and also compensation systems must match the ILMD and choice in SHRM experienced by the firm. In turn, such a match avoids systems that are in conflict with each other. For instance, after Schwab Securities was taken over by Bank of America, its recruitment and selection system was still such that it led to the hiring of individuals who were highly motivated by performance-related monetary and nonmonetary rewards. These employees could, however, no longer be satisfied because the compensation system was slowly but surely adjusted to the ILM rules adhered to by Bank of America.

Compensation systems are especially important to technology acquisition and the necessary organizational adaptations. A compensation system should reflect the firm's strategic push into adopting new technology by including a compensation package and bonus system that rewards appropriately innovative behaviours, or behaviours reflecting favourable attitudes to technology, by employees. Unless the compensation system is, however, consistent with other human resource systems (see also Chapters 6, 7 and 8), conflicts will emerge that can be costly and may lead to antagonism between various parties, hampering the potential success new technology is meant to help achieve.

The next group of chapters illustrates how context and structure as well as strategic human resource management can influence and be influenced by individuals and work-group determinism. Part III highlights how a fit may be accomplished between organizational and individual needs when trying to introduce new technologies into the workplace. A combination of these insights, associated with each of the perspectives in the previous sections of the book, is used in an attempt to account for the emergence of situations in which organizations may appear to do everything possible to facilitate technology acquisition and organizational adaptation while still being jeopardized by individual resistance and work-group values.

PART III

Cultural and Individual Factors Influencing Technology Adoption

So far, this book has used a macro approach to technology acquisition and organizational adaptation. ILMD and choice in SHRM, organizational structure, and processes have been investigated, as has the interaction of HRM subsystems with each other, and the firm's attempt to obtain a fit between these dimensions thereby facilitating technology-induced change.

Up to this point, we have used a contingency approach to describe the functions, processes, and limitations faced by management when trying to introduce new technology. This approach — the underlying premise for all the models we have developed — holds that, for an organization (or its subunits) to be effective, there has to be a fit between its structure and environmental context. This context limits the organization and its managers, designers, and owners in their adoption of new structural designs or the adaptation of the one currently used (e.g., a functional design is changed to a divisional one — uncertain, volatile, and complex environments require "organic," decentralized, and informal structures, and predictable, static, and simple environments call for more "mechanistic," centralized, and formal structures.

To understand fully the organizational adaptation of innovation, we need to integrate the cultural and contingency approaches, which will be attempted in the following part. Combined, the two approaches offer a far more comprehensive model for studying and managing technology acquisition, and its attendant transformation processes, by organizations. Some contingency research has already begun to blaze the trail, so to speak. Miller and Droege (1986), for example, reported that the relationship between perceptions and structure was usually greatest in samples of small and young companies, mostly because the chief executive in such firms wields great authority, and his or her personal desire to achieve (based on his or her own perceptions) might be as influential upon structure as structure is on it.

Part III examines the cultural approach in depth and, more specifically, studies the topics of social comparison and acceptance, matching levels of end-user technology management with skill levels and increasing efficiency with new technology. Environmental determinism, ILMD, and choice in SHRM as well as the components of the human resource strategy (see Chapters 6-9) affect the type of employee working for the firm and help attract a certain type of person wanting to be employed by the organization. Thus acceptance, smoothness of transfer from one type of technology to the next, and utility gains accomplished with the hope of new technology may represent different processes (type, worker participation, and magnitude) by firms in different quadrants. Most important, however, is the fact that all the above are influenced by the cultural context and the environment.

This final part of the book should help to improve our understanding of how people handle organizational adaptation of new technology. We will examine the challenge organizations face in attempting to make their end-user technology investments pay off (in both technological and human assets). The typologies developed earlier can be used to determine some of the characteristics and behaviours exhibited by employees when faced with technology-induced organizational change.

10. The Cultural Context and the Individual

Parts I and II of this book have focused on the objective dimension of technology management, but this chapter concentrates primarily on the subjective dimension. Thus culture (people's attitudes, beliefs, and rituals) is described, and its potential impact upon technology-induced organizational change is outlined. This is not to say that culture itself does not have both objective and subjective dimensions. As Figure 2.1 in Chapter 2 indicates, culture's objective side includes such factors as exploitation of land, organization of industrial work, and laws that can affect technology acquisition and, therefore, must be examined at this point.

Cultures are formed of technical and social artifacts that are cultivated, to a large extent, independently of known and immediately visible reasons. According to Swidler (1986), *culture* can be defined as the publicly available symbolic forms through which people experience and express meaning. It includes beliefs, ritual practices, art forms, ceremonies, and also informal cultural practices such as language, gossip, stories, and daily life observances. These symbolic forms are the means through which the social process of sharing modes of behaviour take place, and we can probably even say that these forms explain most other distinctive behaviours of organizations, groups, and societies as well.

Culture helps shape a *repertoire* of habits, skills, and styles from which people construct *strategies of action*. A strategy is a general way of organizing action (depending upon a network of kin, friends, colleagues, and peers and upon potentially desired outcomes, such as working for a specific employer) that might allow one to reach several different career objectives. *Values* do not shape action by defining its ends but instead help fine-tune the regulation of action within established ways of life, thereby facilitating choice.

Culture and Technology

A culture is stable when the environment sustains existing strategies of action. It becomes unstable when change causes new patterns of action to evolve. Because innovation and acquisition of new technology by their very nature cause such fresh patterns to come into existence, an environment that features an unstable culture is more likely to foster innovation and acquisition of new technology by a firm (Goldstone, 1987). Yet instability alone is not enough to encourage innovation. Goldstone's examination of historical Turkey, China (today's People's Republic of China and Taiwan), and England, as previously discussed in Chapter 1, found that, as far as institutional change went in each of these countries in the seventeenth century, England experienced the fewest changes. The changes going on in that country, however, were sufficiently significant to redefine English culture and encourage (or, at least, not strongly discourage) *cultural diversity*, and this in turn fostered innovation.

More current examples of cultural and political instability are the People's Republic of China (PRC), the German Democratic Republic (GDR), and the Union of Soviet Socialist Republics. The latter's *Perestroika* movement, initiated by its leader Mikhail Gorbachev, is facing opposition as political and economic changes are beginning to redefine Russian culture, advancing less dependence upon the state and more responsibility and decision making by managers and workers instead of central bureaucrats, and encouraging cultural diversity by accepting minorities and various religions.

In the PRC, economic and political change led to political unrest that finally erupted in April 1989 with students demonstrating for reforms including freedom of the press and increased democracy and protesting, among other things, government policies discouraging cultural diversity as a means of controlling developments such as inflation and civil dissent. The demonstration that captured the attention and sympathy of the world exploded on June 3, 1989, when soldiers began killing protesters in Tien An Men Square, the PRC's symbolic seat of power, in Beijing.

On November 10, 1989, the fall's political events in the GDR culminated in the opening of the country's borders, which allowed its citizens free access to the West after decades of severe travel restrictions. A noncommunist president was appointed, free elections are being arranged, and former party bosses are being investigated and jailed for having squandered the people's resources.

There are numerous implications to be considered in these examples. Political upheaval usually affects a country's economy and thus the workplace. The actual process of change also matters as all levels of society must cope with it — willingly or reluctantly. In the case of the PRC, although it seemed that change and instability might lead to an opening of the country to diversity, it led only to chaos. Tolerance for divergence began in the lower levels of society, with the people, but the government "hard-liners" were never committed to the doctrine of change and eventually moved to quash the "rebellion." In contrast, in the Soviet Union, the agenda for change stemmed from the upper echelons but has not been completely successful in sweeping away objections from lower level bureaucrats or Soviet citizens. Without popular support it is difficult to implement change—the miners' strike that nearly brought *Perestroika* to its knees in the summer of 1989 provides ample support for this truism. Still, although it remains unclear whether the changes and present instability in the Soviet Union will prove to be an encouragement or deterrent to innovation, and it is unclear whether the culture will respond to its new potential for pluralism or cling to the secure but stifling status quo, there is evidence that the attempt will continue. One indication is the granting of economic autonomy, beginning in 1990, to the Baltic Republics of Estonia and Lithuania; and Latvia is likely to receive the same privilege soon. The sweeping changes in East Germany are having far-reaching consequences for most of Europe, especially West Germany, where so many East Germans are settling.

Cultural diversity and pluralism are outcomes of positive political shifts. These factors are as important in encouraging innovation in organizational cultures as they are in national arenas. As Perry and Sandholtz (1988) concluded on the basis of an extensive case study, "liberating" organizational forms encourage innovation in firms and facilitate organizational adaptation. Less liberating organizational forms, on the other hand, tend to stifle innovation and its acquisition by a firm. Although cultural instability makes people more readily accept new ideas and innovation, *pluralism* invites people to search for new solutions and strategies and encourages tolerance to cultural diversity (e.g., behaviours, beliefs, and attitudes), thereby facilitating innovation (Goldstone, 1987).

Two contrasting approaches characterize much of the cross-cultural research on organizations. The first is best represented in the work of Hickson and associates (Hickson, Hinings, McMillan, & Schwitter, 1974; Hickson & McMillan, 1981; Hickson, McMillan, Azumi, & Horvath, 1979), who advance a "culture-free" hypothesis and argue that the relationships between certain contextual variables and dimensions of organizational structure are similar across very different societies. This school of thought claims that culture does not affect organizational structures and processes. Specifically, there are (a) positive relationships between organization size and measures of both specialization and formalization and (b) negative relationships between size and centralization for samples of organizations in a large number of different national contexts (Hickson & McMillan, 1981). Miller (1987) used a meta-analysis to test the culture-free hypothesis, examining 23 published papers reporting 27 studies using over 1,000 organizations from 11 different countries. His data confirmed the aforementioned positive relationship between size and centralization, as well as the second point which states that country and type of organization do not significantly affect the relationship between size and centralization of the firm.

The culture-free hypothesis, combined with what we have learned in the preceding chapters, might cause us to believe that cultural differences will not be found when we use the quadrants presented earlier to examine organizations in different nations. This is far from the truth. A second school of thought in cross-cultural research is characterized by a belief that different cultural environments *will* significantly influence the way organizations are structured. For example, Meyer and Scott (1983, p. 14) emphasize the role of *institutional environments* (defined as "including the rules and belief systems as well as the relational networks that arise in the broader societal context") in determining organizational structure and behaviour. A related approach is evident in Child's (1981) discussions of the contrast between capitalist and socialist societies, the cultural differences between Great Britain and West Germany, and the implications of these contrasts for the types of organizational structures and processes evidenced within different societies. Similarly, Maurice, Sorge, and Warner (1980) show that organizational processes develop within an institutional logic that is unique to a society. They believe that organizational structures in a given country will reflect the specific national institutional context more than the effects of "context" variables.

How can these totally different views be explained? It is probable that the proponents of the culture-free hypothesis did not find the kinds of differences that these latter researchers discovered because they were not as particular in their foci, being more interested in overall structures than in more specific processes. Although overall structures and types of formalization do not essentially change across different cultural environments, the internal organizational processes used to adopt technology and adapt the firm, such as work design, training, and decision-making strategies, *do* change (see Miller, 1987). Thus a functional organization or its different levels of hierarchy may be the same across firms in different cultures, but internal processes in these "similar" structures may differ in myriad ways between firms and cultures. For instance, car manufacturers in France and the United States are large firms, hence their levels of specialization, formalization, and decentralization are similar. But we all know that the processes (e.g., from the design stage

The Cultural Context and the Individual   179

of a new car to its final test in the marketplace) of Peugeot and Chrysler differ, and the final products show each firm's characteristic signature.

These subtle differences may not be explicable simply by looking at *objective culture* (which may also partially explain why the culture-free advocates did not notice these changes), such as the man-made environment (e.g., roads, tools, and factories) and level of industrialization. We must also examine *subjective culture*, that is, behaviours, attitudes, beliefs, and values (Bhagat & McQuaid, 1982). The distinction between these terms will be further explained below.

Subjective and Objective Culture

In attempting to define *subjective* and *objective culture*, Triandis (1977, p. 144) suggested first grouping important variables in culture research into several different sets or systems:

(1) the *ecological system*: the physical environment, resources, geography, climate, flora, and fauna;
(2) the *subsistence system*: methods of exploiting the ecological system for survival purposes, such as agriculture, fishing, hunting, and industrial work, including labour laws and employee rights;
(3) the *cultural system*: the man-made part of the environment, which includes both material (roads, tools, factories) and psychological artifacts (norms and values as they exist in the individual's environment);
(4) the *social system*: patterns of interaction, such as social roles, family structure, and institutional behaviour;
(5) the *individual system*: perceptions, motivational patterns, and individual attitudes; *perceived* norms, individual roles, values, and other aspects of subjective culture, which connect the individual system with the cultural system;
(6) the *interindividual system* or *socialization system*: social behaviours (e.g., conformity, helpfulness, aggression, intimacy, covertness), particularly those passed on in various methods of child rearing; and
(7) The *projective system*: traditional myths and fantasies.

Objective culture is usually defined by the subjects of systems 1, 2, and parts of 3, that is, the objective culture of a group might be determined, for example, by topographical factors, food-gathering methods, and degree of industrialization, among other things. Subjective culture, on the other hand, is defined by belief systems, attitude structures, stereotypes, norms, roles, ideologies, values, and so on — the variables classed under the latter systems in Triandis's model. We might also follow Hofstede's (1980, p. 21) lead and define *subjective culture* as "the collective programming of the mind which distinguishes the members of one human group from another." Some of the data for this — beliefs, ritual practices, work forms, ceremonies, and so on — are formally "programmed" into the individual, while the rest enter the psychological data base through informal routes — through "the grapevine," gossip, stories, and daily work experiences. Like computer viruses, data entering through less formal routes may be undesirable in the eyes of the "programmer" (i.e., society), and it is this sort of uncontrolled data that often lead individuals to reject their cultural heritage, much as computer viruses cause programs to crash. Thus formal "programming" affects the individual's value system and repertoire of habits, skills, and styles from which he or she constructs strategies of action and makes choices. As we have noted, the cultural approach is most useful when the contingency approach is baffled by inexplicable differences between seemingly similar organizations. For example, when two organizations appear to experience exactly equal levels of environmental determinism, internal labour market determinism, choice in strategic human resource management, and so on, yet still do not perform similarly, we can deduce some sort of cultural difference. What this difference is may not be immediately apparent, especially if we look at objective cultures, because these coincidentally use predictors similar to those used to study contingency and may, therefore, not appear to vary. Subjective cultures, on the other hand, being determined by an entirely different set of predictors, are likely to provide us with keys to the differences the contingency approach cannot identify. It is mainly for this reason that research using the cultural approach to explain the theoretical processes behind technology acquisition and organizational innovation cannot afford to be ignorant of the complementary results that examinations of *both* objective and subjective cultures might provide.

Subjective culture analysis is especially useful for discovering differences that are *intracultural*, that is, differences *within* what appears to be a distinct culture. Nations, for example, are usually considered to be homogeneous cultural groups, yet there are often enough intracultural differences between smaller groups within nations to make us occasionally doubt the veracity of their homogeneity. These differences are more likely to be between subjective cultures than between objective cultures, usually because most national cultures are identifiable with a single objective culture (though this is not necessarily true, of course, of large countries such as the Soviet Union that are composed of a variety of geographical regions and population groups).

For instance, although the objective cultures of Hispanics and middle-class whites in the United States are similar, their subjective cultures differ because they do not perceive their social environments similarly. The same is true for middle-class Inuit and whites in Canada and middle-class Chinese and Malayans in Malaysia, to cite just a few of many possible examples.

Intracultural differences are not always easily identified, especially when they are based on socioeconomic background, gender, and education, because it is often difficult to see the relationship between these variables and values, norms, and attitudes. Still, the existence of such relationships is unquestionable; we have only to look at the wealth of studies examining differences between male and female norms for evidence of gender-based cultural differences (e.g., Breakwell, Fife-Schaw, Lee, & Spencer, 1987; Gattiker, Gutek, & Berger, 1988; Gattiker & Nelligan, 1988). We are still far from completely understanding why variables such as gender cause intracultural differences, however, and this subject continues to be popular with researchers (e.g., Gutek, Stromberg, & Larwood, 1988; Larwood, Stromberg, & Gutek, 1985).

Cross-National Differences[1]

Cultural differences, as Bhagat and McQuaid (1982) have argued, may reveal themselves in patterns of differences among attitudes. One researcher (Gielen, 1982) found such variations in attitudes in a comparison of ideal self-ratings between U.S. and West German university students. The ego ideals of West German students were those traditionally considered to be "masculine" and reflected a relative emphasis upon wanting to be critical, stubborn, dominating, informed, logical, and foresighted. U.S. students in contrast stressed a desire to be loving, sensitive, perceptive, genuine, and empathic—traditionally "feminine" ego ideals.

Hofstede (1980, chap. 9) found that the values, norms, and beliefs of workers among the similarly structured and widespread branches of a U.S.-based multinational differed according to cultural locales. He was able to distinguish, in each culture in which the organization was operating, unique subjective cultures based on four dimensions: power distance (measure of interpersonal power or influence between the superior and subordinate as perceived by the latter), uncertainty avoidance (tolerance for uncertainty and ambiguity in work roles and one's employment situation), individualism (the degree of individualism/collectivism expected from organizational members), and masculinity (sex differences in work goals—importance attached to advancement, earnings, intersocial relationships, job security, cooperation and training—Hofstede, 1984, chaps. 3-6). In some cases, scores on these dimensions revealed the existence of more than one subjective culture, suggesting intracultural differences: French-speaking Swiss, for example, scored similarly to their German-speaking compatriots with regard to the dimensions of masculinity and individualism but resembled French subjects with regard to the dimension of power distance, and resembled neither of these other groups on tests measuring uncertainty avoidance. This reveals intracultural (within Switzerland) as well as intercultural (between Swiss and French participants) differences.

Because different cultural backgrounds cause organizations to use different methods for adopting new technology, successful international ventures require that the employees of the firms involved make an effort to adapt to (or, at least, come to an understanding of) the cultural strategies of their foreign coworkers. A joint venture between a Japanese and an American firm, for example, would be likely to fail miserably if no attempt was made to indoctrinate employees of one culture with some of the background of the other. Even within one national culture, intracultural differences may make it difficult for employees with different backgrounds within a single organization to cooperate. Blue- and white-collar employees, for example, given that they tend to be from distinct class backgrounds, might approach their work in completely different ways. Once again, the organization will only be able to facilitate adaptation by providing the tools that these employees need to understand and work with one another.

On a national level, German-speaking Swiss (or, for a North American example, English-speaking Canadians) are taught French while the French-speaking Swiss (or French-speaking Canadians) are taught German (English) to help improve understanding of the different cultures and ethnic backgrounds of the two population groups. Nonetheless, knowing a language does not in itself assure that an individual understands the cultural intricacies well enough to facilitate teamwork in a multilanguage and/or culture group. For multinational organizations, it would be most effective to find a culture similar to their own when they wish to establish international branches. For example, a French firm may find it easiest to establish its North American headquarters in Montreal, where most employees are not only truly bilingual but also bicultural (English-French and North American-European). The free trade agreement between Canada and the United States would make such a move even more desirable.

Problems in Studying Cross-Cultural Differences

In the past, organizational research attempting to keep an international perspective has often concentrated exclusively on countries with radically different objective and subjective cultures, such as the United States and Japan (e.g., Luthans, McCaul, & Dodd, 1985), rather than on countries with similar objective cultures, such as the United States and Canada (e.g., Gattiker, 1988b). Although the rationale behind this approach has probably been that differences between markedly distinct cultures are easier to identify, this presents its own difficulties. For example, comparing the organizational commitment of Japanese and U.S. workers is not easy because the cultural framework of Japan instills almost every worker with a sense of commitment to his or her employer that is unlike anything the typical American's cultural background has prepared him or her to experience. Measures of *commitment* developed in the United States, where the word almost has a different meaning, may, therefore, be useless when applied to Japanese subjects, who might even find the questions posed to be inapplicable. If a study examining intercultural differences in commitment or in any other dimension does not explicitly outline the distinct cultural factors that cause such problems, it will be of limited use to anyone who might wish to make use of this information (Bhaghat & McQuaid, 1982).

Another difficulty is the fact that the nations selected may not be similar in their populations' experience with or exposure to attitude surveys. Moreover, social norms governing behaviour related to central topics of the study may differ (Kuechler, 1987). As a result, supposed cross-national differences may be due to these factors rather than to "true" cultural differences. It is easy to see that reported differences in career attitudes between an industrialized country such as the United States and a less developed one such as India could be due to the fact that experience and exposure to attitude surveys is different or due to differing social norms related to work and careers. In such a situation, it may be difficult to explain differences in career attitudes when employing the same research methods in such dissimilar countries (e.g., Sekaran, 1986).

It is especially difficult to identify factors responsible for variations between cultures when there is little similarity between the cultures. This means that, although a comparison of Indian and American bank employees, for example, will most probably reveal significant differences in attitudes (Sekaran, 1986), it will be almost impossible, given that the cultures possess radically different subjective *and* objective cultures, to determine whether one or the other is responsible for these attitudes. Research exploring very diverse cultures is often plagued by this problem. The examination of groups with either a common objective culture (e.g., Caucasians and Inuit in Canada's Northwest Territories) or a common subjective culture (e.g., American and Israeli Orthodox Jews) might be more fruitful, because it would be easier to attribute differences to variations in subjective cultures (as in the first case) or objective cultures (as in the second case).

Admittedly, the study of cultures that are in some sense similar may not provide us with as markedly dissimilar results as the study of very different cultures; in some cases, in fact, it may seem as if the cultures compared are not significantly different at all. However, in such a circumstance, the finer measures used to examine intracultural differences within two countries may find variations (in a comparison of the results from one country with those of the other) that have escaped the grosser measures used in a straight intercultural comparison. By way of illustration, although a comparison of the behaviourial commitment of Americans and Canadians revealed no significant differences, intracultural examinations revealed that income and hierarchic level were more influential on the commitment of Canadian survey participants than they were on U.S. respondents (Gattiker, 1988b), suggesting that there are real differences between commitment in the two countries.[2] The following sections discuss in greater detail how objective and subjective culture may influence the use of technology.

Objective Culture and Technology-Induced Organizational Change

Although this chapter primarily centres on subjective culture, objective culture such as the man-made environment (e.g., roads, industrial work, and laws) can also affect technology acquisition by the organization. For instance, government regulations and incentives may facilitate technology-induced organizational change if the firm is part of a particular industry and located in a remote area. These issues will be further reviewed below.

Organizational Structure

According to the Triandis model described earlier, organizational structures, as man-made artifacts, naturally influence and are influenced by objective culture. Maurice, Sorge, and Warner (1980) discovered evidence that supervisory authority structures differ between objective cultures in a comparison of the organization of manufacturing units in France, West Germany and Great Britain. The percentage of blue-collar workers was by far the highest in Germany and the smallest in France. Total managerial and supervisory staff (i.e., all staff in positions of authority) in proportion to the total number of blue-collar workers was highest in France and lowest in Britain. These authors summarized their work by stating that staff growth had been weakest in the FRG. Moreover, West German firms favoured giving staff positions of direct authority — in contrast to Great Britain, where staff authority was limited (Neal, 1987). In France, staff growth was greater than in the other two countries but balanced between authority and nonauthority positions. The researchers further reported that supervisors are most likely to have authority over production matters in a West German firm and least likely to have such authority in Britain. Friedman (1987) found evidence that culture affects authority structures, claiming that managers in data-processing departments in Japan are more likely to report that salary-related decisions are made outside their department than in the United States, the United Kingdom, Norway, and Sweden; and American supervisors are more likely to note that salary decisions are made entirely within their department than supervisors in any of these other countries.

One explanation for these differences in decision authority could be that they are caused by the various educational systems used in these countries. For example, the FRG attempts to technically professionalize the worker, keeping the qualificational differences between line workers and experts (e.g., technicians and engineers) as small as possible by providing West German workers with a thorough apprenticeship before attaining journeyman certification as well as providing continuous education (company-, government-, or union-sponsored) to attain and retain a high skill level (Casey, 1986). Furthermore, supervisors usually have additional training and certification (*Meisterbrief* or master certificate), which is required for training and supervising apprentices. As a result, the differentiation in authority between line supervisors and experts is minimal (Maurice, Sorge, & Warner, 1980).

Enterprises are generally only willing to undertake general training (procedural knowledge or knowledge about something — its concepts, methods, and theories) if an appropriate share of the cost is borne by the individual. In Great Britain, apprenticeship wages as a percentage of skilled worker rates range from 50% to 100% while in France they start at 80% and end at 100%. In contrast, German apprentices make between 20% and 33% of a skilled worker's rate (Casey, 1986). This illustrates that, from a human capital theory approach, it is not necessarily economical for a French or British firm to provide general training because wage costs are high and the benefits from an apprentice's labour may be minimal or even negative when adding up the training costs incurred by the firm. Moreover, in Great Britain, trades exist mainly in certain traditional manual occupations (e.g., engineering and building and construction industries) and, in France, they are limited to the artisan sector and are, in theory, concerned with providing training for an apprentice to become a self-employed craftsperson. In contrast, in West Germany, apprenticeships exist in all branches of the economy.

The above illustrates that apprenticeships in blue-collar professions are far more important to the economy and industry in West Germany than in either France or Great Britain. This is reflected in having the syllabus federally determined rather than locally or by unions (e.g., Great Britain). The general skill component required for the *Meisterbrief* limits the skill and knowledge differences between apprentices and technicians or engineers, hence, their authority and decision-making power increases. Moreover, the status for journeymen and people with a *Meisterbrief* reduces their desire to become technicians and engineers. In Great Britain or France, status limitations may encourage the individual to become an engineer, and blue-collar workers (e.g., in the metal industry) may not have the same skill levels as their German counterparts, thereby limiting their influence and power in the decision-making process (see Sorge & Warner, 1986).

The political climate is an objective cultural feature that may have a radical effect upon organizational structures. Before World War II, Siemens was based in what is now both West and East Germany. The division of the country following the war made it necessary for the operations of the company's East German plants to be moved to their Karlsruhe facility, with the result that, since the 1950s, Siemens has been based in only three major centres—Karlsruhe, West Berlin, and Munich. Possibly its greatest reason for remaining in West Berlin involves the subsidies (e.g., low corporate taxes and depreciation of capital investments) given to West Berlin-based organizations by the West German government.

Objective cultures often influence organizational structures by affecting the way new technologies that may have an impact on these structures are used. In an examination of how one such technology, computer-numeric control (CNC), affects the shop-floor authority of supervisors and workers, Sorge and Warner (1986) found distinctive cultural variations. West German organizations have used CNC to allow certain decisions to be made on the shop floor and others to be made by designers and engineers, while American firms have used CNC in a fashion that has led to the deskilling of shop-floor workers and to the relegation of nearly all production and design-related decisions to people away from the shop floor (Shaiken, Herzenberg, & Kuhn, 1986). The automobile industry provides us with another example. When Fiat in Italy decided to become one of the most highly automated car manufacturers in the world, it adopted technology that allowed it to centralize its operations, reducing the authority of first-line supervisors and letting top management make the important decisions without much input by workers (Rollier, 1986). In contrast, Volkswagen used similar technology to increase decision making in many jobs and to increase the involvement of workers in the production process (Brumlop & Juergens, 1986).

These examples raise a question: Why does a specific, objective cultural factor cause technology to be used in a specific way? Two factors may explain the differences recorded in the preceding anecdotes. First, high professionalism by German workers and high labour costs may encourage skill upgrading to improve productivity of capital. Second, labour laws require German firms to involve the worker's council in the decision-making process for adopting technology and implementing organizational changes. And, last, job security, based on local laws, is substantial for West German workers. Italy and the United States benefit from lower labour costs but have a less highly skilled work force than West German employers (e.g., Leontief & Duchin, 1986; Sorge & Warner, 1986), making it less profitable to increase worker participation — and the law does not require them to do so. Moreover, the economic situation in Italy (high unemployment) gave Fiat's workers little choice but to accept the changes because resistance could have led to the shutdown of some major plants.

Although objective culture often has an impact on the organization's use of technology, the reverse is also true in that technology may affect culture's influence upon the organization. Information technology, for example, allows for the rapid interchange of information between various offices, lessens the importance of the physical environment in determining the makeup of the organization, thereby reducing the influence of the objective culture to a relative degree. There is no longer a need for organizations to be centralized and to "grin and bear it" when their headquarters' environment becomes unfavourable — through rising real estate prices, for example. Firms now have the option to seek cheap office space outside expensive city centres and to decentralize clerical work (Ellis, 1984). Citibank, for instance, moved its bank card operations from New York to South Dakota and other administrative office operations moved to New Jersey. Harcourt Brace Jovanovich, a U.S. publisher, also moved some of its operations out of New York and presently has offices scattered across the nation (e.g., in Orlando and San Diego), with each handling a different part of its business.

Although technology has helped organizations to better cope with environmental constraints, having offices in different areas of the United States results in the physical environments influencing the makeup of the firm through its influence on subjective culture. For instance, Citibank's employees in more rural South Dakota might have different values (a possible appreciation for outdoor endeavors such as hunting and being involved in community affairs and local sports) from New York employees (a possibly greater focus on indoor activities such as going to the health spa and theatres and attending games played by their local major league team). These examples may exaggerate the differences somewhat by stereotyping them, but, nonetheless, they emphasize that different physical environments become an important factor affecting the people working for the firm, changing the organizational culture and making it more varied from one location to the next. Thus basic assumptions and beliefs shared by employees in one branch may not be shared with their colleagues in another. Of course, such cultural differences might be reduced if branch locations were in either consistently rural or consistently urban areas.

The decentralization methods described in the preceding paragraph, although quite effective in North America, might be less so in a different part of the world where centralization is more explicable by some variable other than geography. And to avoid being misinterpreted as proposing that decentralization is the key to solving the problems of organizations everywhere, we should recognize that, if there is one thing we can deduce from the material presented in this section, it is that differences in objective cultures allow—or actually *force*—organizations to use radically different approaches to increase profitability and productivity, even (as the examples of Fiat and Volkswagen indicate) if these organizations are in similar industries.

Work Flow

Work flow refers to the organization of the transformation process through which inputs such as capital, raw material, and human effort are turned into outputs. This process requires differentiated arrays of machines, jobs, organizational subunits, and technical and social arrangements. Work flow *parallelism* occurs if the same jobs are done parallel to each other and if different parts or components are made parallel to each other. For example, several work groups may install various parts of a car simultaneously using parallel workstations. *Sequential* work flow describes a situation in which a worker performs a specific job on basically the same part, component, information, or other throughput and then passes it on to the next workstation (e.g., assembly line). Thus an individual performs the same task over and over.

The objective primarily affects the work flow through the subsistence system (e.g., the way industrial work is organized). To illustrate, Swedish workers enjoy substantial job security and high wages in comparison with their colleagues in other countries. This forced companies such as Saab-Scania and Volvo (both car manufacturers) to decrease their production costs, which they chose to do by increasing the autonomy of work groups. This helped to speed up production and enhance product quality; simultaneously, worker satisfaction was improved and absenteeism and turnover decreased. The result was that the company's international level of competitiveness recuperated, thus guaranteeing jobs for the employees (Van Houten, 1987).

Manufacturing work flows. Between 1970 and 1980, attempts to humanize work in Sweden took the form of technical control systems designed to improve quality and efficiency and to optimize market-niche opportunities by pacing the manual work of manufacturing work teams (Van Houten, 1987). Although the freedom of workers to organize and pace their work was circumscribed by the input buffers of production (work to be done is temporarily stored in the buffer; once the buffer is full, workers cannot change or interrupt their work pace as incoming work will have no place to be stored prior to being done), work teams were responsible for budget, quality control, and maintenance. Furthermore, job rotation within the teams was done without management's interference. The result was a more flexible production system that enabled the rapid adaptation of the product lines to market-niche changes as well as improved employee commitment to the firm.

In the above example, the force driving work-flow changes was a desire to improve working conditions, to increase the flexibility of the manufacturing process, and to make firms more able to respond to market shifts. Making the work flow parallel by using work teams did not lead to more routinization because team production also involved budgeting, quality control, and maintenance, thus increasing the group's autonomy and responsibility (Van Houten, 1987). This was not so in the case of Fiat, the largest private firm in Italy, which changed the work flow of the shop floor primarily to allow for the use of innovations such as flexible manufacturing, robots, and semiautomated production. The company believed these could increase productivity and reduce costs, allowing the corporation to compete against countries such as Japan, Korea, and Czechoslovakia in the production of low-priced subcompact cars. The work flow remained sequential, the technology took over some hazardous jobs (e.g., painting of cars), and the level of routine was maintained (Rollier, 1986). Organizational structure was changed in that shop-floor workers had little autonomy and practised little job rotation. Despite this, Fiat accomplished its goals of reducing production costs and increasing efficiency, but this was done by *laying off* large numbers of employees who were replaced by the technology.

All three car manufacturers (Fiat, Saab, and Volvo) used technological innovations to change work flows. How could the respective objective cultures help in explaining the different outcomes of technology-induced changes in work flow? Although in the Italian example the methods of industrial production are such that the worker does not participate when decisions are made about job changes, Swedish law requires worker input (directly or via unions). Furthermore, industrial work in Italy is not organized in such a way (e.g., hierarchy, division of labour, and production components) to allow an easy change from sequential to parallel production. Even in its most technologically advanced plant near Cassino, Fiat decided to stay with sequential production while both Volvo and Saab chose parallel production for their new and/or remodelled plants.

Another objective cultural difference could be that Fiat obtained large renovation subsidies for its Cassino plant, because it guaranteed jobs for an economically underdeveloped region of Italy. These subsidies encouraged Fiat to stay with its sequential production strategy, which allows the production of certain car parts in one plant only (e.g., the Cassino plant produces body parts for the Fiat Tipo only), while parallel production favours the manufacturing of a particular car type in one plant. Another important distinction is that the former requires a work flow that resembles the traditional assembly line, but parallel production allows work groups to rotate jobs — they produce whole cars, not just one component (Van Houten, 1987).

In Fiat's case, government subsidies encouraged the firm to increase its production in areas where high unemployment is the norm and skill levels are limited. In contrast, Volvo has a highly skilled work force that values quality of work life and, through its ILMD, has much influence upon the technology-related decision process. A further cultural difference is evident in that, in the United States, factory workers have been shown to be instrumentally oriented to work (viewing work as a means to achieve economic ends; Loscocco, 1989), while in some countries (e.g., Sweden and West Germany), factory workers want to accomplish noneconomic ends (e.g., quality of work life, interpersonal relations and interesting work; IG-Metall, 1984).

Differences in objective culture explain both why dissimilar relationships between work flow and innovation exist and why these relationships were equally successful in enabling companies within various countries to compete in their markets. This seems to be evidence that cultural factors go a long way towards determining the kind of work-flow system that will work in a given country.

Office work flows. Most of the research concerning work flow changes has concentrated on comparing manufacturing processes in different cultures (witness the examples just mentioned) rather than comparing the effects of technology-induced adaptation of work flows in service industries (especially in offices) in different cultures. The success of technology acquisition in increasing office efficiency or reducing administrative costs has so far been limited, however. One reason might be that only processes that are fully understood can be automated or even mechanized (Doswell, 1983), so only simple tasks have thus far undergone this process. Hence, computers are primarily used as support tools by office workers rather than to replace human labour. In some cases, such as mass mailings, the process of duplicating and signing the letter and typing the necessary address labels may be mechanized using the computer, but an employee must still compose the letter and certain commands must be relayed before the computer will start the process of preparing and partially executing a mass mailing. Further change in the work flows involved in administrative and service transformation processes are imminent, however, and are likely to result in productivity gains and cost savings in office settings that will *surpass* those found in manufacturing (Panko, 1984).

Ellis (1984) reported that numerous British firms felt office-oriented technology would enable them to change work flows enough to reduce clerical staff and open decentralized branches. This seems definitely true in the case of the British brewery industry, which introduced office technology to reduce labour costs and to partially automate administrative tasks, control, maintenance scheduling, and management (Davies, 1986, chap. 5). Davies's survey revealed that most of the companies in this industry felt these objectives had been accomplished and that further mechanization and office automation, by changing administrative work flows, would reduce costs and improve their competitiveness even more.

In the British banking industry, both process and product innovations have caused work flows to be rearranged to increase efficiency and product quality (Willman, 1986, chap. 10). Among the more important of these innovations are automated teller machines (ATMs), which enable customers to do their banking at any hour and reduce the need for some clerical and teller staff. In most cases ATMs are a product innovation, offering the client longer banking hours. For the bank, they are also a process innovation, reducing the necessary paperwork so that it does not require a full-time teller but can be handled in one or two brief sessions during the day in the back office.

The rearrangement of work flows in offices often involves a physical rearrangement of the office itself. Offices arranged into open spaces, which allow firms to reduce their office overhead costs, are the current vogue in office design. However, the use of work teams doing parallel work may call for decentralization and correspondingly different arrangements of office space; hence, the open office, or *Buerolandschaft* (office landscaping) concept, may find itself replaced by new concepts such as the *Gruppenraum* (group room). The latter is becoming one of the most popular work arrangements with some groups of West German and Swedish employees (Ellis, 1984), and it represents a movement back to more private office territories where the group controls heat, light, and other services. In North America, however, the move back to more private offices is still in its infant stages and employee groups are not yet pushing for such change. This may in part be attributed to the construction and fixed interior design of high-rise office buildings (see Ellis, 1984), which are far more common in North America than in Europe. These structures, which may not allow for the rearrangement of personnel, represent a cultural factor that affects the North American objective "office" culture as surely as a physical geographic feature like an ocean would and, therefore, limits the potential of North American organizations to use innovative arrangements.

Worker Participation

Although we have already acknowledged the contribution that the acquisition of technology makes towards determining work flows and organizational structures, we have not yet examined how objective culture affects employees' willingness to participate in the process by which technology is introduced into the firm. Employees' participation most likely will be easiest to secure when they are given some voice in the decision-making process. Different objective cultural environments will make this unlikely in some areas and likely in others. In West Germany, for instance, laws outline exactly how workers and their union representatives must be informed about technological changes (e.g., Osterloh, 1986); a similar situation exists in France. It is, therefore, not surprising to learn that Volkswagen has a high degree of institutionalized participation and is quite successful in adopting new technology and making the necessary organizational adaptations. In the United Kingdom, on the other hand, there is a complete absence of a statutorily regulated system of employee representation at the plant and company levels. British unions categorically reject work councils such as those that are well established and have clearly defined rights and duties in West Germany (Neal, 1987). When British Leyland, a car manufacturer, introduced new process technologies in the 1980s, it was able to autocratically impose new work practices, which represented essentially short- and medium-term managerial imperatives to stabilize the company's finances and enhance the government's perception of its future viability, without consulting the unions (Scarbrough, 1986). In the United States and Japan, there also are no legal obligations to consult workers when introducing technological change, yet extensive worker participation is apparent (though the kind of participation varies in each country and in the various organizations). Once again, we must look at cultural differences to explain why these countries do not use the same strategies.

The objective cultural differences described above stem from the labour laws with which the firm is faced. Nonetheless, these objective differences are additionally strengthened by subjective cultural factors such as values, beliefs, and norms that may further distinguish the participation processes used by the three countries mentioned. Hence, even though participation is voluntary in the United States and Japan, the Japanese value changes based on extensive consultation with all parties who may possibly be affected, so technology-induced organizational changes will not be decided autocratically, as is possible in the United States.

Subjective Culture and Technology-Induced Organizational Adaptation

In examining the effects of subjective culture upon the acquisition of innovation, we must not forget that we are much more likely to find differences within cultures and between cultures in the case of subjective culture than we are with objective culture. Therefore, let us look at cross-cultural differences in subjective cultures first and then examine intracultural differences.

Cross-Cultural Differences

Cultural repertoires and strategies. Both a country and an organization's culture encompass symbolic vehicles such as meanings, beliefs, ritual practices, work forms, and ceremonies. According to Swidler (1986), these symbolic forms are the means through which action strategies are shaped. Culturally shaped skills, habits, and styles explain what is distinctive about the behaviour of organizations, groups, and societies.

The action strategies (i.e., one's way of organizing action) and cultural repertoires (habits, skills, and values) that are a part of Japanese management techniques and philosophy, being extremely effective, have naturally been the subject of a large body of research (Tanaka, 1988a), much of it devoted to finding a way to use these strategies in other settings. This may be in vain, however, given that considerable research indicates these strategies may not be applicable in other countries that are quite different subjectively. The ability of the individual to use particular action strategies for managing technological innovation is dependent upon whether or not he or she is able to acquire the new strategies needed to learn the styles of individual behaviour, relationship, cooperation, and so forth innate in the other culture. Thus it is important to assure that the worker has skills, attitudes, opinions, and beliefs to help him or her cope effectively with technology-induced change. Furthermore, if the cultural repertoire needed to implement Japanese management practices is not found in the typical North American or European worker, he or she will not be able to readily use Japanese action strategies (see Hofstede, 1984; Swidler, 1986; Tanaka, 1988b).

The type of action strategies that facilitate individual attempts to rise to the top of the corporate heap appear to differ substantially in countries with discernibly different subjective cultures. In the United States, attending one of the top undergraduate institutions in the country, acquiring an MBA or law degree from a prestigious graduate school, and possessing an upper-class background have all been found to increase the likelihood of rising to the top ranks (Useem & Karabel, 1986). Add to this the fact that, in the past 25 years, chief executive officers with a financial or accounting background have become the dominant group in the *Fortune* 500 companies (Fligstein, 1987), then it is simple to deduce that a prestigious education, upper-class upbringing, and a background in finance or accounting will facilitate learning, understanding, and application of symbolic forms such as beliefs, ritual practices, and values. These in turn help fine-tune the regulation of action within established ways of life and propel individuals up the corporate ladder. In France, by contrast, certain *Grandes Ecoles* (elite graduate schools in public administration with powerful alumni) facilitate advancement in public service and in private industry; attendance at one of these is considered to be far more important than the actual type of degree obtained. In Japan, we find an entirely different situation again; experience in various departments (e.g., marketing, production, and accounting) is as important as education, and both are used as screening devices for promoting individuals (Tachibanaki, 1987). The well-bred American trying to rise in a Japanese corporation would, therefore, be more likely to find that his or her pedigree was useless.

Given that the executives of an organization in one culture will probably have used similar tools in their climb to the top, they can be expected to have comparable outlooks. This implies that establishing a joint venture between two firms from dissimilar countries may be very difficult because the outlooks of the two groups of executives involved may be quite different. American executives, for example, commonly having financial or accounting backgrounds, may look at technological innovation from a short-term financial gain perspective, while Japanese executives may expect a long-term return on investment, having a better understanding of the whole organizational picture and of the effects of new technology acquisition on different departments. A look at the Japanese and American steel industries shows this to be so. Japanese producers introduced new technology based on product and process considerations even though the financial benefit was hard to establish with any accuracy at the time. American producers, not seeing any immediate return on their investments, did not adopt the same technology, which may in part explain why they have lost much of their market share to Japanese competitors (Gold, Rosegger, & Boylan, 1980).

Joint ventures between countries with radically different *political* systems present even bigger challenges for economic success. Production output (e.g., quantity, quality, design, and type of product) may predetermine the degree to which the demand for the product can be satisfied and also the level of consumer satisfaction in a country like the Soviet Union or the PRC. In the West, of course, consumer demand determines both the type of production and the kind of product that is generated. This has made recent joint ventures between companies from the West and East difficult to implement. The limited decision-making powers of the manager of a state-owned enterprise and the notorious unreliability of suppliers of raw materials and other production components in the East are considerable obstacles to overcome when a joint venture based in an Eastern-bloc country attempts to apply the Western philosophy of supply and demand. The habits, skills, and management styles of state-enterprise managers will have to change if the enterprise is to compete effectively in its new business environment.

Worker participation. German supervisors, as Sorge (1985) showed, are more likely to use participatory approaches when introducing changes in the workplace than the antagonistic methods preferred by their British counterparts. The theory that objective cultural differences give German supervisors far more decision authority than British supervisors might in part account for this (Maurice, Sorge, & Warner, 1980), but more subtle differences in subjective cultures may also provide a rationale. It is possible that British workers refuse to cooperate with the firm in the decision-making process, not only because there are no laws giving them these rights but because to do so would be to go against the principles of the "class struggle" in which they believe themselves engaged (see Ford's negotiations in 1988). The subjective cultures of workers and management in Britain cause them to be far more conscious of class distinctions than Germans and thus unwilling to admit that workers and management have common interests. This is manifest in the antagonistic "reactive" conception of trade unionism in Britain, which makes it difficult to consider any collaborative action or participation in decisions affecting the future well-being of the organization (Neal, 1987).

Different subjective cultures will naturally cause different attitudes towards such things as class struggles. In Germany, to contrast with Britain, a different conception of social class causes the work force to perceive employers as *Sozialpartner* (social partners) rather than as nemeses, and there is more belief in the theory that the organization and the work force are working towards common goals. Subjective culture may be the prime mover in encouraging the establishment of laws requiring the formalized participation of employees or their representatives in the decision-making process in some countries or (as is probably the case in Britain) in discouraging these same laws in others.

Subjective cultural differences in the kind of participation that the organization can expect from its employees means that the multinational must make allowances for the particular conditions in each of the countries in which it operates. For example, although Japanese multinationals use similar management philosophies and organizational structures in their headquarters and in their Canadian subsidiaries (Jain, 1985), they respond to subjective culture differences in these subsidiaries by "Canadianizing" their management style — that is, changing their human resource and labour relations management strategies. Among other things, this means that quality control circles, common in Japanese plants, are far less common in Canadian subsidiaries, in recognition of the fact that Canadian employees are reluctant to participate in the organization after working hours without pay (Jain, 1985). The attitude towards participation that is peculiar to the Canadian worker is also one of the reasons why bottom-up communication, prevalent in the parent organizations, is less common in Canadian subsidiaries, where communication tends to be mostly in the other direction. Japanese managers are conditioned by their subjective culture to feel that change and initiative should come from those close to the problem. Canadian subjective culture, on the other hand, conditions managers to receive orders from above and equates authority with hierarchic level. This discourages employees from caring about improving work problems, and they are less likely to perceive their own welfare and that of the organization as being intertwined.

This illustrates that multinational firms must consider and adapt to local culture. Although adjustment to objective culture may be relatively easy (work laws, transportation, and ecology), it might be more difficult in the case of subjective culture (beliefs, habits, values, and attitudes).

Risk management and technology acquisition. The operation of any sort of organization is by no means a sure thing; there is always a certain degree of uncertainty, which is especially true for the organization that wishes to stay competitive through the adoption of innovative methods. The type of risks the organization will experience can be divided into two kinds: systematic and unsystematic. *Systematic risk* emanates from *overall market risk* and cannot be diversified away (Bettis, 1983). Typically, risks caused by changes in the overall economy or by periodic occurrences such as tax reform fall into this category (Hogarth, 1980). *Unsystematic risk* is unique to the firm and is associated with the organization's specific resources and competencies as well as its relationship with the environment. New technological developments and striking work forces represent unsystematic risks.

Because the "growing pains" associated with the introduction of any new technology are often severe (Gold, 1983), management usually puts increased emphasis on quantifiable results, in which case decisions concerning the actual adoption of innovations are based on whether or not set goals can actually be determined (e.g., O'Reilly, 1983). This process can be complex (Driver, 1979), so it is not surprising that individuals tend to simplify the given problem. Cost preference behaviour — that is, analyzing transactions by focusing on cost efficiency — is popular in the area of human resources (see Walker & Weber, 1984). Assessing unsystematic risk in SHRM is accomplished by measuring behaviourial outcomes such as employee turnover and absenteeism in economic terms (Cascio, 1987; Mobley, 1982). Another way of reducing unsystematic risks is for the company to participate in risk-sharing by entering joint ventures (common between manufacturers and retailers) or by arranging for centralized activities such as warehousing (Spulber, 1985).

The above techniques are especially predominant in North America but are also emulated to a large extent elsewhere. Differences in risk management may occur. Because the subjective culture of American managers causes them to focus on costs, their planning is often influenced by projected quarterly results and by the price of their company's shares on the market (Arnould, 1985). This short-term financial focus may in turn make the unsystematic risk factor that is inherent in the acquisition of new technology appear too great to be desirable. Because this is the general trend in American business, very successful companies, such as WordPerfect Corp., that do not subscribe to this philosophy may be reluctant to go public out of a fear that investors may force them to abandon their long-term management focus. This would prevent them from taking on unsystematic risk or introducing technically advanced products before their competition—the very methods by which these organizations became successful in the first place (A. Ashton, personal communication, April 9, 1988). The long-term perspective often held by European investors, in contrast to the short-term perspective characteristic of Americans, reduces the pressure upon management to produce high profits on a quarterly basis. Similarly, the high banking debts of Japanese multinationals reduce similar pressure because bankers are more interested in long-term performance on their loans than on quarterly results.

Subjective culture, as we have noted, tends to determine the outlook of managers, and it just as certainly influences the outlook of the organization's investors, thus determining the kind of short- and long-term perspectives they tend to have. Because these outlooks decide the organization's approach to risk, subjective culture can be said to indirectly determine approaches to risk management. The differences caused in outlooks by various subjective cultures mean that approaches to risk management will vary from culture to culture. Some cultures, therefore, are more likely to use approaches that facilitate technology-induced change than others.

Intracultural Differences

Because intracultural differences may exist between the sexes, racial groups, religious sects, and so on within a larger culture, employees from different classes or groups within a country construct their action strategies from different repertoires of habits, skills, and styles. These differences, according to Bhagat and McQuaid (1982), are often not fully explained in research.

Gender. How subjective culture causes women and men in the workplace to differ in their relationships to innovation in general and to computerization in particular has received little study (Form & McMillen, 1983). Traditionally, research comparing women and men in the work force has been primarily interested in comparing them with regard to the leadership domain, promotion and income patterns, and job satisfaction (e.g., Larwood, Stromberg, & Gutek, 1985). In addition, an extensive review of the literature on women and work indicates an interest in discovering gender differences based on workplace segregation and family interdependence (Stromberg, Larwood, & Gutek, 1987).

In North America, women are affected more frequently by computerization than men because they are more likely to hold occupations and positions compelling them to work with this technology (Form & McMillen, 1983; Gutek, 1983). Although, in rare cases, females do feel more positive than males concerning computerization—for example, they are more positive about computer technology's benefits for white-collar employees (Gutek & Bikson, 1985)—studies examining their attitudes have generally revealed sentiments that were more negative than those expressed by males, even in studies that attempted to include respondents of both sexes with similar jobs and responsibilities (Gattiker & Nelligan, 1988). These feelings are caused by a number of different fears regarding the effects of this kind of technology. In some cases, there was a belief that computerization would reduce career possibilities and the quality of job life (Gattiker, Gutek, & Berger, 1988; Gutek & Bikson, 1985). In Canada specifically, female respondents were less positive about computerization than men because they foresaw decreased communication and increased control, while in the United States, by contrast, female respondents were more worried that computerization would decrease their quality of job life (Gattiker & Nelligan, 1988). The variations between Canada and the United States are due either to differences in women's roles in the respective work forces (e.g., Larwood, Stromberg, & Gutek, 1985) or to different subjective cultural experiences in the two countries.

A study by Gattiker and Hlavka (1989) reported that Canadian female university students enrolled in microcomputer courses did not differ much from males in their attitudes towards technology and, in fact, were more positive than any other group when they owned computers themselves. Level of education alone, however, does not seem to be enough to change negative attitudes; female British university students had less positive attitudes towards technology than males (i.e., being less likely to aspire to jobs in high technology fields such as engineering and research; Breakwell, Fife-Schaw, Lee, & Spencer, 1987). It seems that a specific kind of education, one based on the technology itself, is required. Similarly, the fact that women are more often brought into contact with technology through their work yet still remain less positive about it suggests that it is the kind of exposure rather than the amount that affects attitudes. Because it is also likely that the female students who owned computers had positive attitudes to begin with, one could hypothesize that a willingness to learn, instilled through the educational process, would make the amount of exposure more worthwhile.

Hierarchic level. Some North American literature indicates that values and beliefs held by employees can differ based on their positions within the corporate hierarchy. For instance, Useem and Karabel (1986), who studied top management positions in the 100 largest U.S. companies, found that these employees' socioeconomic backgrounds (father's profession and position as well as family income) instilled certain values (need for prestige and power) that ultimately propelled their careers to the top of their respective organizations. McClelland (1975) already reported that managers tend to have higher needs for achievement and power than their nonmanagerial colleagues. As such, managers might be motivated to increase control over their subordinates' performance, thereby facilitating their own goal pursuit and career achievement.

The above line of reasoning is substantiated by data gathered in the United States that also suggest that computers are perceived by employees as a method of increasing managerial control (Gattiker & Nelligan, 1988), which in turn requires the managers to have knowledge of the technology. Moreover, computers have become a tool for North American academics, who use them extensively to type papers and books and do statistical analyses. In Austria or West Germany, the hierarchic structure is such that the professor rarely uses a computer and instead has his or her assistants and secretary(ies) do these "chores."

The above indicates that status thinking could in large part explain why certain groups of employees may resist new technology, especially if its use would lower one's status (e.g., typing). Thus, in North America, computer skills are expected of a manager who considers him- or herself to be up to date (e.g., Kaufman, 1989), while in German-speaking countries and most of Europe, computer skills are not considered a necessity. Such differences are, however, primarily based on values and beliefs about managers' jobs and responsibilities.

Other cultural factors. Youth is fashionable in North American society. Not only did the major 1989 movies feature female stars with an average age of 29 years (Ansen, 1989) but most advertising features youthful figures. This youth orientation goes so far that, in supervisory ratings, a slight tendency to rate older employees lower in their performance is apparent (Waldman & Avolio, 1986). Apparently, supervisors might automatically place older employees into categories that are unrealistic—assuming, perhaps, that they are likely to tire easily, "lose their competitive edge," and so on. Performance ratings also showed more positive relations with age for professionals as compared with nonprofessionals—again, possibly the result of a categorization system that classes older professionals as "wise grey heads" and older nonprofessionals as "tired old men."

Other societies place a greater value on older employees. The Japanese assume that younger employees require a substantial amount of experience before they can even be considered for supervising others (e.g., Tanaka, 1988b). Similar reasoning leads the Swiss Bank Corporation to have its Swiss foreign exchange dealers be in their midthirties before obtaining a certain amount of responsibility and power. In contrast, their U.S. branches have

The Cultural Context and the Individual 201

some dealers in their early twenties. Again, cross-national differences in values and beliefs led the bank to have different career paths for their Swiss headquarters and U.S. branch employees in the foreign exchange department. This is not the place to judge these different approaches but to point out how cultures can differ in how they feel about age and how this can affect corporate career paths.

Another factor that may affect technological change is religion. For instance, if their religion forbids men and women to work together in the same room or if prayer is required several times during the day, work flow and technology use must allow for the necessary interruptions. Also, on "high holy days," work is not permitted regardless of whether it is technologically feasible, economical, or desirable (see Tanaka, 1988a).

In a U.S. study of religious beliefs in the work force, Chusmir and Koberg (1988) found that persons without religious affiliations had a higher need for power than those who reported being Protestants, Catholics, or affiliates of Eastern religions. Furthermore, managers with religious convictions had a higher need for achievement than nonmanagerial personnel with religious convictions, suggesting a connection between work attitudes and religious convictions. Therefore, religion, as it affects the workplace, can be a double-edged sword, either hindering or aiding technology-induced organizational adaptation.

Different technologies and technology attitudes. Different groups of employees doing similar work with various technologies often differ remarkably; microcomputer users, for example, are more often satisfied with the quality of their work lives than users of mainframe terminals (Gattiker, Gutek & Berger, 1988, using a U.S. sample from *Fortune* 500 companies). Although a different study, using a slightly changed scale and surveying workers in companies of various sizes, did not have exactly the same results, the responses of Canadians surveyed showed that people working with personal computers felt more in control and thought that communication opportunities (within and outside of the firm) were better than those working with a mainframe terminal (Gattiker & Nelligan, 1988).

Penrose and Seiford (1988) did a study comparing two popular personal computers (IBM-PC and Apple Macintosh). The two computers were perceived as having distinct personalities by the participants of the study. For instance, the Macintosh was perceived as being easier to use, more flexible and nontechnical than the IBM-PC by both IBM and Macintosh users. According to the authors, these results suggest that different hardware and software designs and marketing strategies may in turn affect users' technology perceptions.

Similar occupational groups and their technology attitudes. Different occupational groups perceive computer-related technology in a variety of ways, a phenomenon that is apparently tied to exposure to the technology. Zoltan and Chapanis (1982), examining the attitudes of CPAs, lawyers, pharmacists, and physicians towards computers, showed that CPAs had the greatest amount of training and exposure to computers while lawyers had the least. Given this latter fact, it is not surprising that lawyers were also most likely to describe computers with negative terms, such as *depersonalizing, formal,* and *difficult.* Pharmacists emphasized terms like *adaptable, fun, personalizing,* and *enjoyable* to describe computers and CPAs stressed the functional attributes of computers with terms like *powerful, helpful,* and *affordable.*

The above occupations use computers in officelike settings for informational purposes. The differences reported in attitudes may in part be due to different knowledge levels held about the technology (see also Chapter 12) and the extent of exposure to the technology. Technology attitudes seem to become similar between occupations if the amount of knowledge based on training and use is also similar. Thus an engineer using a computer-aided design (CAD) system might have different attitudes from those held by a lawyer (see Klein, Hall, & LaLiberte, 1990).

Different occupational groups and their technology attitudes. Breakwell, Fife-Schaw, Lee, and Spencer (1987), surveying a British sample, reported that students with occupational aspirations towards technical professions had more positive technology attitudes than their peers. A position that requires working with computers all day is more likely to increase stress and negative health effects than one using the technology as a support tool (Stellman, Klitzman, Gordon, & Snow, 1987). Hence, secretarial personnel may feel less positive about computer-mediated work than engineers and lawyers (Bikson, Gutek, & Mankin, 1987; Morgall, 1983). Additionally, job segregation has resulted in secretarial positions being generally held by women (e.g., Bielby & Baron, 1986), which may explain why women often feel less positive than men about computers.

The above indicates that different occupations have different attitudes about computers and technology in general. Once again, this could in part be explained by the use of technology. Mechanization of certain tasks in the office has hit women harder as the majority of clerical and secretarial personnel are female. Thus clerical personnel may not feel as positive about technology as managers do. The latter may in turn use the technology as a support tool while nonmanagerial personnel use the technology to perform most job-related tasks.

Summary

The research shows that a clear distinction must be made between objective and subjective factors that can affect technology acquisition and organizational adaptation either separately or in tandem. The discussion indicated that the way technology is integrated into the production process may differ significantly based on objective and subjective culture (Willman, 1986). Concentrating on both types of culture may prove to be the most satisfying plan for the firm as well as its employees.

As we have seen, although laws and regulations can drastically affect the extent to which the work force participates in the technology acquisition and organizational adaptation process, the values, norms, and behavioural strategies used by employees and management are the deciding factors in determining the content of the participation process. As Wilpert's (1986) study showed, laws in themselves may not guarantee worker codetermination and participation in the early planning process for new technology. Furthermore, changes in office design and other areas, such as electronic banking, in various countries will ultimately affect work redundancies in the service industries.

The presence of numerous cross-cultural and intracultural differences restrict the manager's ability to use a generalized strategy for technology acquisition in different cultural contexts. A strategy that works for Japanese employees will, therefore, not necessarily work for American employees, and a strategy that works for Japanese males may not work for Japanese females. Limited knowledge about the impact of certain technologies on individuals or the work environment may also result in resistance that will vary according to the group expressing it, thereby limiting the potentially positive outcomes of using new technology.

It is impossible to be concerned with different cultural groups' attitudes towards technology and not become concerned with whether or not questions concerning equal worth affect these attitudes. More and more attention has been focused on the equal worth debate, which has become proportionally more intense as rapid technological change has redefined work roles. Because inequality may greatly relate to technology-induced organizational changes, the following chapter addresses problems of equal worth.[3]

Notes

1. The terms *cross-cultural* and *cross-national* are used interchangeably in this book. In the technology context we are especially interested to see how national cultures (e.g., the United States versus Italy) influence technology acquisition and technology-induced organizational change.

2. Whereas two nations (e.g., Canada and the United States) are similar in language and ecology, differences in their subjective cultures would result in differences in their belief systems, attitude structures, stereotypes, norms, roles, ideologies, values, and task definitions. Thus differences obtained for such constructs as organizational commitment must be explained by investigating the subjective cultures of the countries included in a study. When differences on many of these constructs fall into a pattern, we can identify genuine cross-cultural differences.

3. Recently, new laws have stated that jobs that are of equal economic worth must carry equal salaries. For instance, nurses in Washington state filed a law suit against the government because they felt that their work was of equal value to that performed by the maintenance workers in hospitals (mostly men) but their salaries and fringe benefits were significantly lower. The court decided in the nurses' favour.

11. Inequality and Technology Acceptance

Chapter 10 illustrated how important cultural factors can be in determining technology acquisition and technology-induced organizational change or adaptation. Culture also influences the attitudes held by the individual about new technology, and cultural systems can affect the inequality experienced at work by such varied minorities as women, immigrants, AIDS victims, and natives. Furthermore, the inequality experienced can be perceived differently in various countries based on the victims' socialization, which may make such inequality acceptable or unacceptable for the individual. We are interested in how these cultural factors may interrelate with the social comparison process — people comparing themselves with their colleagues in similar work situations and drawing conclusions — and also how this might affect the technology acquisition process and potential success of technology-induced organizational change.

This chapter concentrates on the individual (micro) and subjective dimension of our model (see Figure 2.1, Chapter 2). How does culture affect perceptions of inequality and the social comparisons that influence these perceptions, and how will these elements in turn affect employees' acceptance of new technology and the organizational adaptation process? For brevity's sake, the chapter highlights inequality by using women as one example of a minority experiencing inequality in the workplace as well as inequality's relationship to acceptance of or resistance to new technology.

Internal labour market determinism as outlined in Part II of this book may sensitize employees to how technology-keyed changes affect equality and equity by causing them to compare their own situations with those of peers and friends in the firm and in other organizations. As we saw in Chapter 9, the acquisition of new technology may change job descriptions and classifications, which in turn might adversely affect compensation. The latter part of this chapter then outlines how the model of organizational types developed earlier can be applied when considering technology acceptance. We will specifically discuss how organizational types, based on their distinctive cultures, differ in managing technology-induced organizational change by looking at the approach management takes toward inequality caused by this kind of change and the way in which their work forces will probably respond to such change.

National Culture and Inequality for Women

Women earn less than men. Even after adjusting for differences in levels of productive resource endowments, substantial residual disparities remain in earnings of demographically comparable working men and women. Past attempts to determine why this discrimination continues to exist in an age of supposed increasing sexual equality, have focused mostly on establishing how much the male-female salary gap in organizations may be *directly* due to differences in rates of return on stocks of human capital (wage discrimination) or *indirectly* created by inequalities in the distribution of men and women across job classes and hierarchic ranks (segregation; e.g., Halaby, 1979; Kemp & Beck, 1986; Wagner, Ford, & Ford, 1986).

Human Capital Explanations for Inequality

The general theme pervading the human capital literature is that individuals should be financially remunerated on the basis of their productivity or job performance. The latter is assumed to be largely determined by formal schooling and special skills. Later developments in human capital theory also emphasize variation in postschooling investments, in particular, tenure in an organization and in a position (Mincer, 1974). Consequently, greater variation in work, organizational, and positional tenure among job incumbents should result in greater intraoccupational earnings dispersion.

However, formal schooling and job experience may benefit women less than men in terms of compensation (Halaby, 1979). Such gender inequality leads to "unequal pay for equal work," with the male-female pay differential exceeding the corresponding productivity differential due to discrimination (Becker, 1971; Kemp & Beck, 1986).

Institutional Explanations of Pay Differentials

This theory proposes that wage differentials are due largely to disparities in the allocation of jobs and promotions, which results in segregation along sexual lines, with women crowded into the lower-paying and lower-productivity positions. Although a considerable body of research examines the effect of different industries on individual earnings (e.g., Davidson & Reich, 1988;

Hodson, 1983; Parcel & Mueller, 1983), there has been little speculation about particular industries' effect on inequality between sexes. It has, however, been argued that increasing numbers of public employees reduce overall economic inequality because the government sector has the least variability in wages and salaries of all industrial categories, largely because of political forces that help keep earnings similar in public bureaucracies (Kemp & Beck, 1986).

Cross-Cultural Differences and Similarities.

Inequality based on gender does not stop at national boundaries; the work forces of most industrialized countries experience it (e.g., Cooper & Robertson, 1986). Although this inequality is attributable to similar factors in most of these countries — usually, to the culture's assumption that a woman's place is in the home — its magnitude is affected by the diverse value systems of these cultures. This explains why gender-based inequality is less extensive in Scandinavia than in the Mediterranean countries, where *machismo* is a much more tangible feature of the cultural environment.

In Japan, despite an equal employment opportunity (EEO) law that went into effect in 1985, women must still cope with gender discrimination in hiring, promotion, and job assignment, and certain jobs are not open to women at all (Edwards, 1988). In the Japanese case, it is quite simple to identify the two general factors that limit women's potential for hierarchic and lateral career movement (Edwards, 1988). First, Japanese women generally start their careers in clerical and secretarial positions with limited advancement opportunities. Second, many Japanese companies base the promotion process in large part on organizational tenure (Tachabinaki, 1987). Women, because they often leave the work force sometime during their childbearing years, have less tenure and are not offered jobs higher up the ladder when they eventually return to the organization. These factors, unfortunately, inhibit the potential of the EEO law to reduce gender-based discrimination in the work force to the extent the lawmakers surely intended (Edwards, 1988).

Being female, as the Japanese example suggests, may often determine the kind of position an employee might hold. A North American study by Bielby and Baron (1986) also determined that some positions are segregated — that is, women don't usually hold them and, when they do, their job titles are different and allow for lower remuneration. Their study was based on data from numerous organizations of various sizes in California. In one case, men who were employed by a pharmaceutical company were titled "senior aerosol packaging *technicians*" while women doing the same work were called "aerosol packaging *operators*."[1] Apparently, not all sex segregation within organizations and within occupations in the United States is caught by the three-digit occupational codes used in the *Dictionary of Occupational Titles*.

Organizations often justify paying women less by citing the fact that they have less formal education than male employees; in a Swiss study, approximately 40% of differences between women and men were attributable to women's lower level of education (Kugler, 1988). Formal education in itself does not necessarily make one employee a better performer than another with an informal education, yet this is often the excuse given by firms who discriminate against women wanting to enter male-dominated professions. As in North America, it might be necessary to let the courts decide if such a selection criterion is discriminatory unless it can be proven that without the educational credentials, satisfactory performance cannot be attained.

At any rate, despite what employers might argue, education may not be the answer for women seeking to rise within the corporation. According to Lorence (1987), who used an American sample, increased education is not a tool that can help propel a woman to the top as it does a man. At the same time, increased education does not make it as likely for women to have higher salaries as it does for men. Lorence reported that, in his sample, males still receive higher earnings than women within the same occupation who have similar educational backgrounds. Lorence tested human capital as well as institutional determinants and found substantial support for the institutional determinants once again, thereby indicating firm-initiated discrimination against women. These findings are supported by Larwood and Gattiker (1986), who reported that higher education did not relate as significantly to occupational status for women as it did for men. Institutional factors (e.g., letting women perform jobs without giving them the appropriate job title) may in fact prevent women from profiting from their advanced education as much as men do when progressing through the corporate hierarchy (Gattiker & Larwood, 1990).

Lorence (1987) also reported that his data demonstrated considerable economic inequality among females in the same occupation. In his female American sample, more years of schooling tended to decrease overall intra-occupational economic inequality among women but had no significant effect among men.

Economic Development and Technology-Induced Organizational Change

Women in Finland often benefit financially when they perform the same sort of work as men because male-dominated occupations tend to be better paid (Kauppinen-Toropainen, Kandolin, Haavio-Mannila, 1988). This increase in monetary compensation was more apparent for white-collar than for blue-collar women workers. For men, however, the effects of segregation on job characteristics and occupational prestige were the opposite. When men enter traditionally female job areas, wages overall usually increase.

Most new technologies have the potential either to improve or to worsen working conditions for both women and men, but a review of the literature on technology and women indicates that new computer-based technology mostly affects women in clerical positions (Gutek & Larwood, 1987), possibly because this technology can be used most effectively to mechanize and automate relatively simple and easily understood tasks that are often the province of these positions (Doswell, 1983, chap. 1).

Gattiker (1989) compared Canadian with U.S. respondents working with computers in office settings. He reported that Canadian women experienced more wage discrimination than their U.S. counterparts. Moreover, extensive use of computer-based technology seemed to increase the level of wage discrimination to the disadvantage of women in both countries.

The above is, of course, not only limited to women. It seems feasible to assume that certain other minority groups experience similar developments. Nonetheless, research is rather scarce and none could be found dealing specifically with technology, inequality, and minorities based on race or ethnic background.

Even when inequalities based on gender or race are present and identifiable by outside observers, insiders may not perceive them (Znanieck Lopata, Fordham Norr, Barnewolt, & Miller, 1985). It is, therefore, important to discuss further how individuals may interpret and assess their work situations and to try to determine any effects on their perceptions of inequality.

Technology Acquisition and Inequality: Social Comparison by Employees

Several theoretical conceptualizations of equity in organizations, most notably equity theory (e.g., Adams, 1963), have focused on how the distribution of forms of organizational compensation, both material and cognitive, affect attitudes and perceptions (see Greenberg, 1982, for an extensive review). Although this legacy of theory and research has provided a great deal of information about reactions to the nature and level of organizational compensation, most of it does not attempt to explain what causes employees to perceive inequalities in the first place; neither does it explain why apparent inequalities may occasionally fail to affect emotional processes (Landy, 1985; Scholl, Cooper, & McKenna, 1985). *Social comparison theory*, on the other hand, suggests that social comparison information might be the key to this problem. According to this theory, individuals selectively use comparison information to learn not only about themselves but about the social environment (Festinger, 1954). By comparing him- or herself with others in higher positions and to those with similar demographic characteristics (e.g., education and gender), the individual learns what is "standard" in the environment and is then free to judge how his or her own situation compares with the standard (Goethals, 1986). The individual also uses this method to determine reactions to his or her own state. If an employee perceives that he or she is being treated unequally, yet sees that there are others in the same situation who do not let this treatment affect their work, he or she will be likely to follow suit and continue to work without complaint. Although social comparison theory apparently helps explain why individuals react differently in cases of inequality, it is still not entirely understood. Little is known about how individuals actually select the information they use in comparing themselves with others (Wheeler, 1986).

Points of Reference for Comparison

According to social comparison theory, different types of employees use different points of reference as guides for their actions. Women tend to compare themselves with peers of the same sex to determine how to react in a given circumstance. If their colleagues do not fare any better than they do, there is a great chance that they will not, as a result, feel unfairly treated (e.g., Gattiker, 1988c). Individuals with substantial organizational tenure use information about the job complexity and compensations of their coworkers as points of reference when evaluating these job facets (Oldham, Kulik, Stepina, & Ambrose, 1986). In one study, 75% of a sample of employees used points of reference for job evaluation purposes, and more than 60% compared their jobs with more complex positions (Oldham et al., 1982). Managers, on the other hand, as Goodman (1974) has shown, often use themselves as well as others as reference points when evaluating pay. Such *self-comparison* involves a comparison of one's current situation with past or (hypothetical) future situations (Adams, 1963); many people, for example, assess their financial success by looking at their past, current, and possible future life-styles (see Gattiker, 1985b). Incidentally, employees who use themselves as a reference point for comparison have been shown to be more likely to leave an organization than their peers (Oldham, Kulik, Ambrose, Stepina, & Brand, 1986).

Gender. The carryover of gender-based expectations (e.g., the woman is the homemaker and nurturing parent) into the workplace may influence the comparison process. Women, for example, despite the results of a social comparison, might not feel deprived or less satisfied with a situation if some sort of justification prevents them from perceiving inequality. For example, Crosby's (1982) study that surveyed American women from the upper socioeconomic strata found that respondents felt as satisfied with their work and jobs as men, even though they earned less and inequality was apparent. Two justification factors are at work here. First, most of the women in this study were comfortably well-off without the money they received from working; according to Shepelak (1987), they would thus be likely to find pay inequality easier to accept. Second, these women usually found their jobs to be interesting and challenging. Together, these two justification factors cancelled out the ill effects of wage inequality. If, on the other hand, the factors used for justification were insufficient — say, the job was boring and the employee was desperate for money — she would, according to the *insufficient justification hypothesis* proposed by Pfeffer and Lawler (1980), evaluate the situation negatively.

It is well established that victims of structural inequality often fail to challenge socioeconomic injustices and, in many cases, support the very systems that keep them disadvantaged. A recent study reported that, although women experienced wage discrimination as well as job segregation, they did not feel less successful in their careers than men (Gattiker, 1989). Inequality may also be due to differences in expectancies with respect to various job outcomes among female and male job applicants (Hollenbeck, Ilgen, Ostroff, & Vancouver, 1987). Bigoness (1988) claimed that, when it comes to professional growth, female MBA students have higher attribute preferences and lower preferences for salary development than male MBAs. In yet another study, women showed less concern with pay than men but equal job satisfaction (Crosby, 1982, p. 161). By taking full responsibility for their relative attainments, having lower expectancies, and putting more emphasis on nonmonetary rewards, disadvantaged people may come to believe they have no cause to protest inequalities in the distribution of extrinsic rewards (e.g., compensation and promotions).

Cross-cultural similarities and differences. Objective differences between the work situations of employees of one gender in one national culture (e.g., the United States) and employees of the same gender in a different country (e.g., Canada) do not necessarily lead to altered perceptions of inequalities. Even though Canadian women may experience higher wage discrimination than their U.S. counterparts, Canadian women may not feel any less successful in their careers than U.S. women (Gattiker, 1989). This could in part be explained by the fact that socialization processes for women may differ between the two countries, preparing women for different life roles. Canadian women may be acculturated to perform more of a traditional, supportive role than their U.S. counterparts. Furthermore, U.S. women have started to enter male-dominated professions in far greater numbers than Canadian females, thereby changing their society's perception of women's work roles. The result will probably be that the female who appears to be unequally treated from a comparative cross-national perspective may not feel that way, in part due to socialization and cultural upbringing (see Crosby, 1982).

As described earlier in this chapter, Gattiker (1989) found that wage discrimination was higher in Canada, and Canadian women did feel less financially successful in their careers than Canadian men. Thus intracultural differences (Canadian females versus Canadian males) in career success perception do exist although intercultural differences (Canadian females versus U.S. females) did not materialize. Of course, because Canadian women have continual contact with their male countrymen and considerably less, as a rule, with their American female counterparts, their tendency to use Canadian men as a reference point is readily explicable. Canadian females may also have become sensitized to discrimination, a supposition borne out by their starting to challenge it in various ways (changing employers, going to court, and initiating strikes[2]). In many other countries, such as Japan and France, this sensitizing process has, however, not yet progressed as far.

Another issue may be the cross-cultural differences between occupational groups based on attempts by unions to help women in reducing wage discrimination or job segregation or, conversely, their failure to do so. To illustrate, ILMD and industrywide labour contracts may reduce objective inequality in some countries because remuneration packages may in part be determined by job classifications. This clause can be circumvented, however, by not giving females the same classification as men doing essentially the same job. Most unions have not been eager to tackle this issue, thereby leaving women with limited institutional power and resources to change such inequality in a firm. However, the fact that many women accept inequality due to socialization is crucial; they are too used to playing second fiddle to men in out-of-home activities such as work. Unfortunately, there is as yet little cross-cultural data about inequality in the workplace or, most important, about how this may affect women's perceptions and attitudes towards work and careers.

Technology and perceived inequality. Technology-induced organizational change, as we have noted, affects women more than men by mechanizing or eliminating many of the simple tasks for which they are most likely responsible (Doswell, 1983, chap. 1; Form & McMillen, 1983). This usually leads to further simplification and specialization in a cycle that limits the advance of women in the organizational hierarchy even more (Morgall, 1983). Thus it is perhaps not surprising that females are more apt than males to feel that computer-based technology will hinder their careers and reduce their occupational status (Gutek & Larwood, 1987). Neither are the results of Gattiker and Nelligan's (1988) study unexpected, which found that women in both Canada and the United States are more likely than men to believe that computers affect quality of job life negatively. As this demonstrates, technology is likely to cause perceived inequalities not only in earnings but also in the desirability of job characteristics.

It is also important that we determine further how social comparison processes, social information, and the individual's own assessment of the situation lead to resistance to change due to technology acquisition and organizational adaptation. For instance, it appears feasible that a previously unsuccessful introduction of new technology can lead to future resistance against other technology by employees. The next section deals with the construct of acceptance and its importance in the technology acquisition process.

Acceptance of Technology

Psychologists often try to assess *technology acceptance* using a cognitive perspective and, therefore, interpret it as an evaluative attitude (see Oskamp, 1977, chap. 1) towards a new technology. Technology acceptance is, of course, affected by the cultural environment. If a society perceives a technology in one way, the individual's own construct will be similarly influenced to some degree, and he or she may be socialized to react to a certain kind of technology without ever having encountered and experienced it. The Bavarian peasant who condemned the evil Baron Frankenstein's mad scientific experiments even before the resulting monster ran amok and destroyed his cottage was probably more influenced by the society around him than by his own rational approach to the subject.

In a survey of the literature dealing with resistance to computerization, Dierkes and von Thienen (1984) defined a hypothetical construct called "acceptance of computerization" to measure employee reactions to, and beliefs about, computers. They suggested that overcoming the resistance of particular employees to changes caused by computerization is greatly facilitated by the overall public acceptance of computerization itself. Unfortunately, their literature review also shows that the way the individual arrives at his or her beliefs and attitudes about computers, which is to say, how social comparison affects the individual's construct of acceptance, is poorly understood at this time.

Because perceptions and attitudes towards technology affect behaviour (for example, acceptance should lead to a positive behaviour such as effective use of the technology), *acceptance* as used in this text combines the psychological with the behavioural approach. Acceptance or resistance is, therefore, an attitude formed towards technology by an individual, based on two components. The first element is *determinism*, which is grounded in social information processing theory and argues that individual attitudes are not a function of deep-seated needs but a product of how people socially construct the world around them (Staw, Bell & Clausen, 1986). Hence, the *social information* concerning new technology, its effects, and the individual's interpretation of this information (perhaps provided by others or through the observation of one's own behaviour in its social context) influences his or her level of acceptance. The second component affecting acceptance is *social comparison*, through which the employee attempts to determine what his or her attitude towards computerization should be on the basis of how others react to it. Social comparison differs from social information mainly by having the individual specifically compare his or her own work situation and technology-mediated work with others, possibly those in higher positions or with similar demographic characteristics. Based on the above elements, the individual forms his or her attitude which may or may not be conducive to an acceptance of technology acquisition and technology-induced organizational change.

Acceptance of Technology Acquisition

Figure 11.1 illustrates the resistance/acceptance continuum. If an individual starts off with a low level of resistance towards technology, possibly caused by a concern about its effects on his or her job, additional training and information might help him or her move towards a higher level of acceptance, depending on the effectiveness of the training and the individual's own personal makeup. If an individual shows a high level of resistance before the introduction of new technology, he or she will probably be much less likely to accept the technology later on, although it is probable that his or her resistance will be reduced after some time if he or she wishes to remain employed (Gattiker & Larwood, 1986b).

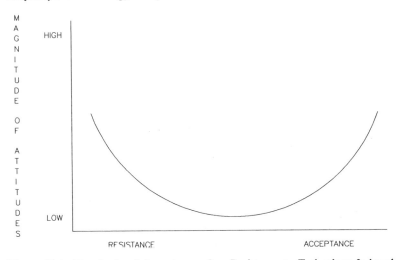

Figure 11.1. Magnitude of Acceptance of or Resistance to Technology-Induced Organizational Change

The attitudes of individual employees may, indirectly, have a decisive impact on a firm's possibilities for technology-induced organizational change when expressed through the group voice of the ILM or a union. In instances of high ILM determinism, therefore, attempts to adopt new technology are greatly facilitated when employees have positive dispositions towards such change. A high level of resistance originating in a highly structured ILM may effectively sabotage the organization's efforts to make changes. It is basically for this reason that we must study the factors that influence employees' acceptance and rejection of new technology.

Ergonomics. Ergonomics is research that studies the design of machines, products, workplaces, and jobs with the purpose of improving worker productivity and well-being. Ergonomics involves taking the physical and psychological characteristics of human beings into account when designing and implementing new technology, the arrangement and structure of the work environment, job organization, and so forth. Both unions and organizations have accepted ergonomics as a way of making work more human (that is, less dangerous, less monotonous, and more interesting, challenging, and satisfying) and, at the same time, increasing the work output (in quantity and quality) and reducing costs related to accidents, health problems, and turnover.

Ergonomic research often reveals information that contributes greatly to our understanding of workers and their attitudes towards new technology (Long, 1987, pp. 270-280). For example, ergonomic studies have reported that computer technology may be responsible for everything from physical impairment in hands, arms, and shoulders (caused by keyboard construction and placement) to visual problems (associated with great contrasts in luminescence between the screen and the source documents, as well as with high oscillation degrees of characters; Fellman, Bräuninger, Gierer, & Grandjean, 1982; Newton, 1984). There have also been complaints of significantly higher levels of job and physical environment stressors in full-time VDT users when compared with part-time ones, such as typists and clerical workers who do not interact with machines regularly (Stellman, Klitzman, Gordon, & Snow 1987). The kind of problems associated with the technology listed here, because they are perceived to be directly linked to the technology itself, can lower acceptance levels greatly. Because complaints tend to increase with the amount of time spent using the technology, the firm wishing to implement it might be best advised to limit the amount of time employees spend with it.

Quality of technology-mediated work. Technology-mediated work will be perceived as lowering the quality of job life if the public views it as being less prestigious than other work, and this will cause employees to resent the technology (e.g., Petermann & von Thienen, 1988). Information systems, although they are used in more and more organizations, are the targets of extensive criticism. The quality of the information these systems provide their users is often seen as substandard (Nutt, 1986), and, although studies have asked managers to formulate their information needs, in the hope of solving this problem, the information needs of support personnel, who use the technology in far greater numbers, have been neglected (Mankin, Bikson, & Gutek, 1984). The manager who makes every effort to know the needs of his or her subordinates with regard to information systems will be able to ensure that the technology is used to its full potential (Gold, 1983; Nutt, 1986).

One of the most common drawbacks of innovative new technology is that it often makes work more abstract and presents a barrier to employees' comprehension of the processes involved in their work (Sydow, 1985). Statistical software, for instance, has become very user-friendly. It can be used by employees who have little knowledge about the assumptions made or the defaults used by the program when calculating the statistics—who are thus unaware of their effect upon the type of results obtained. Thus less conservative defaults (e.g., replacing missing values with mean values obtained) may translate into unjustifiably risky decisions although more conservative defaults (e.g., ignore cases with missing values) could prevent this. In both cases, the results may vary substantially and lead to different decisions. Knowing something about statistics would probably allow the individual to program the software to make more conservative assumptions, resulting in more justifiable decisions with acceptable levels of risk.

Task identity — the degree to which the job requires completion of a whole and *identifiable* piece of work — can also be affected by technology acquisition and organizational adaptation. The employee can more easily identify with his or her job if involved in tasks from the beginning to the end and when some kind of appreciable outcome is in sight. New technologies, by making work more abstract, have a way of diminishing or even destroying task identity. Often an employee may not see the relationship of his or her work to a finished product or may fail to understand how the computer processes and distributes information. If this happens, employees will perceive quality of work life to decrease.

Job impact. The impact of computers and information systems on work prospects and job security are widely discussed in the literature. There is little agreement on whether or not further developments in office technology will help create new jobs for workers who find themselves already rendered obsolete by such technology (Gutek, 1983). A belief that computer technology, and especially information systems, will have a great impact on the individual's employability is often the driving force that encourages him or her to acquire computer literacy (Gattiker & Larwood, 1986b; Lippitt, Langseth, & Mossop, 1985, chap. 1). Knowing whether or not their employees possess these beliefs about computerization, therefore, becomes important to managers who want to see the technology in which they invest used correctly (Card, Moran, & Newell, 1984).

Technological transformation is likely to entail a large number of more or less simultaneous changes. From the employees' point of view, it is highly unlikely that these changes will only affect a single aspect of an individual's job, and it is even more unlikely that all those individuals affected will view the consequences in the same light. Based on these individual differences in perception and interpretation, some employees will define themselves as "winners," making them most likely to accept the technology. Others will define themselves as "losers"; these workers will probably resist technology-induced changes. Roskies, Liker, and Roitman (1988) reported that self-perceived "winners" are strongly convinced that change, in the form of automation, is a necessary condition for company survival. And although the process of change entails a certain amount of current disruption — assuming new jobs, learning new skills, or even simply working harder and longer — "winners" consider this a fair exchange for improved future job prospects with the firm or elsewhere.

"Losers" are remarkably similar to "winners" on all points (relationship to firm and personal characteristics) except the crucial one: They feel excluded from the new company and the new society. Automation- and technology-related training constitute the ark built by Noah as protection against the rising waters, but those who consider themselves "losers" feel they missed the boat. Signs of this situation are the uncertain relevance of their jobs to the company's new mission and the absence of specific job-related technical training given to these employees by the firm (Roskies, Liker, & Roitman, 1988). In contrast, "winners" get job-related technology training by the firm and see the technology acquisition as a positive event, necessary to assure future survival of the firm. Presumably, individuals perceive themselves as "losers" based on the information and signals they receive, such as lack of technology-related training offered to them by the firm and uncertainty about job security and future career prospects. Hence, a firm is well advised to inform its employees far in advance about how technology-induced changes may affect them in their jobs, thereby preventing negative attitudes from developing.

Facilitating Technology Acquisition and Technology-Induced Organizational Change

Even though individual assessments of the potential effect of technology upon job-related duties may affect employees' willingness to acquire or improve their computer skills, the question remains on how to ensure their technology acceptance. Some research claims that people do not tend to change their beliefs (Staw, Bell, & Clausen, 1986). Yet, if they believe that computers will have a significant impact on their current and future work situations, rational individuals may come to accept computers as necessary evils, but without changing their beliefs (see Larwood, Gutek, & Gattiker, 1984).

To facilitate the introduction of technology, *first*, it is necessary for the organization to assess the beliefs held by its employees. Based on this assessment, the organization should then begin an information campaign designed to alleviate any concerns and promote the expected changes. This might involve training employees in the use of the new technology and providing them with an understanding of its capabilities and limitations. It is also important (as discussed earlier) that the organization try to offer job security and absorb projected redundancies via attrition, outplacement, early retirement, and retraining.

Technology and inequality. The above illustrates that technology acceptance is in part dependent upon the organization ensuring that ergonomic concerns and job impacts are also considered. Furthermore, management must ensure that inequalities do not arise due to technology-induced change. If they do occur, the employee may determine through social comparison and social information processing that the job situation has deteriorated.

Although new technologies may make the skills needed to operate older technologies obsolete and thus enable the firm to hire cheaper, less skilled workers, a powerful ILM will usually demand that the firm continue to pay veteran employees at a higher rate, regardless of the changes in job content. The compromise that typically results is the two-tier wage structure discussed in Chapter 9. Although such structures may seem attractive to veterans and to the organization, it is unlikely that they will be as popular with newer employees, who will naturally compare themselves with their more experienced peers and find their own situations wanting. Even though many might think that the ILM or the organization itself should be the target of employees' anger, many employees will actually blame the technology that caused the formation of these structures. This is likely to happen whenever any inequality between employment situations can be traced back to the introduction of some kind of technology.

The social information the employee receives may also be traced to occurrences in different work settings or different employers. Such information may convince the individual that a potential technology effect could be lower quality of work life or job redundancy. Again, such information can lead the employee to resist the acquisition of technology in his or her current workplace. On the other hand, social information may cause a different employee to believe that having the new technology at his or her disposal is a symbol of status.

The inequality issue has also gathered some considerable attention in public since the effect of computerization upon jobs held by women has been addressed in various countries. For instance, a study reported that, in West Germany, the positions that would experience the greatest degree of additional mechanization and automation in the coming years were held by women (Krebsbach-Gnath et al., 1983). A study by Ruch and Troy (1986) that concentrates on the effect of computer-based technology upon secretarial work in Switzerland again shows that women have to cope with radically changing working environments, in large part due to technology acquisition by their firms.

The above indicates that computer-based technology particularly affects clerical jobs held primarily by women. Possible negative effects such as deskilling may, therefore, largely be experienced by women. Based on social comparison, however, such outcomes may be perceived as increasing inequality (wage, job segregation, job impact, and quality of work life), resulting in increased technology resistance. Thus the challenge for firms is to make sure that technology outcomes are not perceived as further increasing inequality.

As a continuous process, the acceptance of technology-induced changes in the workplace can only be assured if the firm performs periodic assessments and adjustments accordingly. Over time, numerous side effects and less desirable outcomes in the workplace may be blamed on the technology and thus reduce acceptance levels (Gattiker, 1990b). It is, therefore, of some interest to management to ensure periodic data collection and its subsequent discussion with employees and occupational groups as well as unions. Only such a participatory approach will provide management and employees with the necessary information to reduce anxiety and the negative work effects caused by technology and continuous change, assuring an acceptable quality of work life for the employee.

Organizational Culture, Technology Acquisition, and Acceptance

An important factor affecting the individual's acceptance of technology acquisition and organizational adaptation is the cultural stability he or she experiences. A relatively high level of cultural stability increases the potential for resistance and prevents individuals from adopting the cultural values, strategies, and habits needed to cope with instability due to technology-induced organizational change (Swidler, 1986).

Although organizational culture and the human resource culture in particular are important factors influencing the acquisition of new technology (as discussed in Chapter 4), the national culture of which the organization is a part also affects the way technology acquisition and organizational change will be handled. For instance, a culture that encourages change may allow firms to depreciate new technology faster than a country less tolerant of new ways (Gibbs, Keen, & Lucas, 1988). Furthermore, tolerance towards change may also encourage governments to establish funds that facilitate the retraining of workers with obsolete skills. Even though firms may be in similar quadrants, based on their geographic location and predominant local cultures,

Inequality and Technology Acceptance 221

they will differ in myriad ways when it comes to handling technology acquisition and technology-induced organizational change. This interrelationship between organizational type and culture, as well as technology acceptance and inequality, is outlined below.

Quadrant A

Firms in Quadrant A are usually bureaucratic and feature socialistic human resource cultures (see Table 5.1). The organizational culture is stable, that is, basic assumptions and beliefs are shared by all members and have been in place for a considerable amount of time. Because the compensation system reinforces these beliefs, organizational members tend to resist change. Labour relations policies typically focus on maximizing the number of employees. The job redundancies, content changes, and reclassifications caused by the firm's acquisition of new technology are, therefore, likely to be resisted by employees, placing the organization in this quadrant in an especially difficult position. Technology-induced change, if accepted at all, must be compensated for by guaranteed internal equity. This will make it hard for the organization to pay the going rate for rare skills as this would cause salary increases across the board.

In this environment, organizations must be prepared for the resistance of employees to new technology, a resistance that is not grounded so much in a general refusal to work with new technology as it is in the fact that employees do not really have the necessary skills and strategies to cope with change (Perry & Sandholtz, 1988). Stability of organizational procedures and policies does not enable employees to adjust easily to changes in the structure or in interdepartmental and interpersonal relationships caused by the introduction of new technology. These factors make continuous, step-by-step adaptation of organizational processes more likely to succeed in this kind of business instead of radical or discontinuous organizational change and acquisition of technology (see also Chapter 4). Unfortunately, however, the very conservativeness that characterizes such a firm usually causes it to delay the introduction of change as long as possible, until such time as it is forced, by market pressures, to discontinuously introduce radical new innovations (see Table 3.1 in Chapter 3).

Quadrant B

High ILMD encourages a more bureaucratic culture with a formalized HRM strategy focusing on cooperation. Organizational change and the acquisition of new technology is more common in this quadrant. The company will try to offer job security and deal with potential job redundancies by offering workers retraining, early retirement, and, if necessary, outplacement counselling. Although the organizational culture is relatively stable, some diversity (e.g., values, beliefs, and action strategies) does exist, which facilitates organizational adaptation somewhat.

Inequality due to technology acquisition may result in further wage discrimination and job segregation for women (e.g., clerical positions are affected by computerization and subsequent deskilling may lead to lower remuneration). Management will consider both internal and external equity when choosing labour contracts and remuneration packages; external equity, for example, may dictate that employees with certain job classifications are paid more than others because they possess scarce skills, while internal equity will ensure that the differences will be kept at a minimum.

Employers in this quadrant might experience some initial resistance from employees to the introduction of new technology. However, as these firms do enjoy choice in SHRM, technology acquisition is backed by a long-term strategy permitting an evolutionary introduction of new technology (see Table 3.1) and proactive change (e.g., retraining, transfers, attrition and outplacement counselling). Furthermore, the high ILMD gives employees the assurance that management must keep them informed about such changes before they are introduced. Thus, over time, these individuals will tend to move toward acceptance of new technology.

Quadrant C

Because the ILM culture of the Quadrant C organization is more participative (see Table 6.1 in Chapter 6), and because the organization is a technology prospector (see Table 3.1), technology is usually introduced in an evolutionary manner even though the innovation may be considered radical. The "winner" attitude of the organizational culture and the high quality of the compensation system make it viable for the company to encourage continuous change. To facilitate such change, the company allows for some cultural diversity and considers external equity when drawing up labour contracts. Employees with complementary skills are paid more and employees with rare skills are paid the market rate or better for their services.

Nonetheless, because ILMD is low, the firm may allow inequality to increase due to technology acquisition, unless obvious wage discrimination and job segregation are against the law. Competitors who tolerate less inequality, and a sparse labour supply, may prove to be even more powerful forces for change. However, this only applies for highly skilled employees;

for low-skilled manpower, the only help in reducing such inequality would come from the law.

In this quadrant, individuals perceive the acquisition of technology and organizational adaptation as a challenge. These employees tend to consider themselves "winners" and see such technology acquisition as the edge that allows the company to remain competitive, which also makes their jobs secure (Roskies, Liker, & Roitman, 1988).

Quadrant D

In Quadrant D, the human resource culture of the organization tends to be apathetic and labour relations are either antagonistic or indifferent (see Tables 5.1 and 6.1). Stability and conformity are encouraged, and cultural diversity is kept at a minimum. Inequality due to technology acquisition is generally not due to management's conscious attempt to accomplish this outcome but due to default. Hence, the firm does not necessarily increase inequality because this might improve profitability prospects but simply lacks a strategy that would take advantage of the situation.

From an outsider's perspective, the firm does not appear to have a coherent strategy regarding the inequalities that might be caused by technology-induced change. Neither an internal nor an external focus is the driving force for reducing inequality. Instead, it may often depend upon the employee or a group of workers to raise the issue with management and insist on change (e.g., salary increase).

Employees in this sort of organization may eventually accept almost any type of new technology as long as it does not cause too much disruption and change. Unfortunately, indifferent acceptance of the technology will probably mean that it is rarely used to its fullest potential (see also Chapter 12) — and it may lead to exploitation of the workers by management. Only limited provisions are made to account for technology-induced job changes and interruptions in the work environment. Firms situated in this quadrant fail to take advantage of low ILMD by not having a clear strategy for their HRM — they are reactors when adopting new technology.

Summary

It is unlikely that inequality in the workplace will ever become totally extinct. Gender-based inequality is well documented and occurs in most industrialized countries. At the same time, however, this may cause some serious difficulties for two reasons. First, because women tend to hold lower-level jobs whose processes are mostly understood, their tasks tend to become mechanized first. Hence, women appear to be affected by technology more than men and the outcome is more often perceived negatively. Second, more and more women have joined the work force, and single-parent families headed by females have become a substantial percentage of the total number of families. As a result, women need wages that allow them to satisfy their families' needs — the old myth that women provide only supplementary income is rapidly becoming obsolete.

The first point may result in women resisting technology as much as possible, endangering its effective use and undermining the justification for the huge capital outlays, yet the second point will result in women forcing unions to address this issue or taking their disputes to court. Apart from ethical considerations, which in themselves require management and unions to strive to prevent this scenario, such outcomes reduce the benefits society hopes to gain through technological advancement. Inequality (objective as well as perceived) needs to be reduced for women if technology acquisition by organizations is to become completely successful.

New technology may also affect minorities differently (e.g., blacks in the United States, natives in Canada, guest workers in West Germany and Switzerland, and certain castes in India). Unfortunately, research dealing specifically with this issue was not available. Nonetheless, technology's effects upon these groups must be assessed in the future.

This chapter also suggests that successful technology-induced change requires all workers to be allowed to participate in the change process, with job security to be ensured if at all possible. Thus redundancies should be managed by offering retraining, outplacement, and/or early retirement options to employees (Brumlop & Juergens, 1986). Specific job-related technical training needs to be offered to all workers so they will feel like part of the technology acquisition process rather than like helpless victims of a company, which is in turn equally constrained by its environment.

Looking at the organizational types discussed here (Table 11.1), it is obvious that any organization, in any quadrant, will experience some inequality, which may in part be caused by technology. A firm must, however, make a serious effort to reduce its effects. Failing to do so may force the firm to cope with increased resistance. Table 11.1 suggests that firms in Quadrants A or D are most likely to have a work force that will try to resist change. Cultural stability in those firms will make it difficult for the employees to cope with technology-induced change in their work. Most important, how-

TABLE 11.1: Organizational Culture and Technological Adaptation

Variable	Quadrant A High ILM Determinism, Low Choice in Strategic Human Resource Management	Quadrant B High ILM Determinism, High Choice in Strategic Human Resource Management	Quadrant C Low ILM Determinism, High Choice in Strategic Human Resource Management	Quadrant D Low ILM Determinism, Low Choice in Strategic Human Resource Management
Organizational culture and change	Job redundancy, job content, and classification	Job security and avoidance of redundancy	Challenge, appreciation of cultural diversity	Stability and conformity
Stability of culture	Stable	Some diversity and change	Diversity and continuous change	Generally stable
Inequality due to adoption of new technology (e.g. wage discrimination and job segregation)	Occurs but is not apparent at first glance	Does occur in some instances	Occurs quite frequently	Average occurrence
Management focus if technology-induced inequality occurs	Internal	Internal, possibly external (mostly legal considerations)	External (competitors and legal considerations)	No coherent strategy/focus
Employee acceptance of change	Perceived as threat, resistance high	Some resistance, but more likely moving toward acceptance over time	Perceived as opportunity and challenge	Resistance or acceptance but most likely indifference if job security is assured

ever, it is likely that these firms will have to introduce new technology radically (see Chapter 3) and discontinuously, because they are usually reactors (see Table 3.1). This in turn will further strengthen resistance. For these types of firms a shift toward another quadrant may be the only way to survive in a more competitive marketplace with a public sensitive to the inequality issue.

As outlined earlier, acceptance of new technology and the necessary organizational adaptations to make effective use of it require a comprehensive strategy in HRM. SHRM should particularly help in developing a work force whose computer literacy or technology "know-how" fit the level of end-user computing or technology-mediated work attained in the organization. Acquisition of new computer-based technology will ultimately affect the job content, skill levels, and other organizational processes, and it may also cause or increase inequality. One strategy for management to use to reduce inequality and ensure effective use of technology is training its work force, enabling workers to acquire the newly required skills needed to succeed in a high-technology environment and giving them the confidence that they can do so.

This chapter has outlined how the subjective dimension—the individual (micro) level of analysis—must be considered when trying to determine a firm's effectiveness in the technology acquisition and technology-induced organizational change process (see Figure 2.1 in Chapter 2). Moreover, objective factors such as ILMD, HRM systems, and policies (e.g., compensation and career development) are ultimately closely linked with perceived levels of inequality and social comparison processes (see Table 11.1).

Notes

1. In the province of Ontario, Canada, hospitals had hired men as cleaning technicians (full-time, seniority rights, and fringe benefits) while females were employed as cleaning personnel in part-time positions doing the same job-related tasks as their male colleagues. In early 1989, a provincial court decided that this situation discriminated against females and thus it forced hospitals to pay back wages and fringe benefits to females. Furthermore, hospitals were ordered to offer employment to female cleaning personnel under the same conditions as male employees (i.e., full-time and same job title, wages and fringe benefits).

2. On September 5, 1989, Quebec nurses started an illegal strike triggered by the provincial government's unwillingness to increase their wages by 20% across the board in 1989, thereby reducing the alleged wage discrimination experienced by nurses compared with other health professionals in the province (e.g., lab technicians and maintenance personnel). The government penalized nurses by reducing their wages by two days for each day they were on strike and, by

September 8, the nurses started to lose one year of seniority for each day they were on strike. Moreover, the government hired replacements and took the union leaders to court for organizing an illegal strike. This example indicates that legal channels must be used to reduce wage discrimination by taking employers (in this case, the provincial government) to court to enforce employment equity. Legal strikes may give workers the opportunity to improve current working conditions and wages but may fail to reduce wage discrimination for a variety of reasons.

12. Employee Skills and Technology Management

The end-user domain has become one of the most interesting areas in the study of technological innovation and technology-induced organizational change. The end-user is usually an office or production worker, teacher, tool-and-die maker, ship's mate, or a doctor (to list a few) rather than an engineer, programmer, or computer scientist who is making use of advanced technology (computer-based or other) in his or her work.

Use of advanced technology by such a diverse group of employees with various backgrounds and know-how requires the firm to establish end-user technology management (EUTM). EUTM naturally has the end-user at its nucleus and considers quality of job life, job security, and occupational health to be major issues (see Stellman, Klitzman, Gordon, & Snow, 1987). EUTM could be defined as a system of interrelationships between financial, technical, societal, and human constraints, which must be carefully balanced to achieve equilibrium and to assure *constructive use of the technology*. In a more practical sense, constructive use of a technology is accomplished only when satisfactory levels of job security, quality of work life, health and safety can be maintained for the employee; when ecologically justifiable environmental impacts and ethically acceptable use of technology are assured; and when levels of productivity and profitability attained are such that make the use of advanced technology feasible. All stakeholders, therefore, find it in their own interest to constructively use technology in the production process.

The spread of technology in the production process has increased the interdependence of subsystems and processes. To illustrate, if a computer-based airline reservation system (see Chapter 1 – SABRE from American Airlines) is down due to system maintenance or power failure, travel agents are no longer able to make airline, car, or hotel reservations; if a welding robot in a computer-integrated manufacturing system breaks down, production assembly must be stopped. To minimize the ramifications of such technology breakdowns, employees must have the skills to effectively cope with these situations to minimize the damage.

Technology breakdowns caused by faults in computer programs, wear and tear, or human error are unavoidable. Employees' coping skills, or lack thereof, can increase or decrease the magnitude of the damage caused to the environment, human life, and other species as well as to the firm. For instance, the environmental disaster caused by the oil tanker Exxon Valdez along the Alaskan coast in 1989 was in all probability caused by human error in using the ship's advanced navigation technology. Moreover, skill levels of cleanup crews were inadequate to cope with the oil spill efficiently, thus unnecessarily multiplying the damage done to the coastline and marine life. Similarly, the Chernobyl nuclear plant accident of 1986 in the Soviet Union was caused by human error and imperfect technology. Human, animal, and plant life was severely affected within a large geographical zone, necessitating the evacuation of more than 100,000 individuals. Long-term genetic effects will not be known for many years. Moreover, radioactive fallout was carried by winds, contaminating vegetable and other crops as far away as France, Sweden, and the United Kingdom. However, if the Chernobyl plant employees and the disaster rescue teams had acted more efficiently, the damage would have been contained. Workers' skill levels must, therefore, enable them not only to apply the technology in their work, *but equally importantly, first, to anticipate and prevent technology malfunctions or breakdowns if at all possible and, second, to cope effectively with accidents if the technology breaks down.* Consequently, necessary skill levels and appropriate decision making should have told Chernobyl personnel to shut down the overheating nuclear reactor on time, which could have prevented the accident. Better understanding of the ramifications of their negligence — more knowledge of the damage a nuclear accident could cause — might have prompted workers to learn and remember steps needed to contain the damage in an emergency.

In light of these examples, it is obvious that EUTM must in large part deal with the skill issue. Unfortunately, the current skill debate either rarely addresses the technology issue directly (e.g., Spenner, 1983) or focuses exclusively on the effects upon quality of work life as manifested by upskilling and deskilling effects attributed to technology (Attewell, 1987b; Braverman, 1974). In this chapter, the skill dimension is considered to go beyond the worker's or firm's well-being (i.e., quality of work life, job security for the employee, and profitability for the firm). The interrelationship of skills with environmental factors, such as educational systems and safety issues, are also discussed, and, in contrast to previous work, the focus is primarily on the office and not the factory.

There are two reasons for concentrating primarily on the office setting. First, technology-induced organizational change and advancement in the office may be the greatest challenge for the next decade (e.g., with computers). Furthermore, the potential for cost savings and increased productivity appear to be greatest in the office (Panko, 1984). Second, the office setting has been neglected by researchers, making our knowledge about technology's effect upon office work rather limited (Gutek, 1983).

As earlier chapters have noted, the specialization of certain occupations in some countries may have reduced general knowledge, thereby making it more difficult to upgrade the skills needed to perform well with more advanced technology (see Chapter 10). Moreover, based on contingency theory, it could be argued that the environment and, especially, the market the organization faces may influence EUTM and skill levels attained by employees (see Chapters 2-4). For instance, a fast-growing market represents continuous change (e.g., culture, product innovation, market entry of new competitors), and new technology may penetrate the whole organization and be at a higher level than technology in a firm operating in a stagnating market. To cope well with a growing market and its resultant environmental changes and innovation, the firm requires a work force willing to adapt and learn new skills continuously. In a static market, technology is also necessary to reduce production costs (personnel, operations, materials, and asset management) by, for instance, speeding up the work process. Negative outcomes of this process, from an employee's perspective, could include more structured work and a simplification of tasks leading to possible deskilling.

To keep our discussion simple, a brief overview of the literature discussing skills in conjunction with technology is given, followed by a conceptual model that combines psychological and sociological thought in a new approach. Using this model, a conceptual scheme outlining skill effects based upon the firm's quadrant location is presented (see Figure 3.1).

End-User Technology Management and Employee Skills

Effective EUTM requires the firm to deal with the issue of employee skills. Technology-induced organizational change may alter the present skill mix substantially. Consequently, technology acquisition can only be handled beneficially for all involved if employee skills and inherent job skills can be matched successfully (see Schein, 1978).

It would be useful to provide a simple definition of *skill*, but the term is not one that is easily defined; indeed, analysts have been struggling with its meaning for decades (see Adams, 1987, for an extensive historical review). Recent theories of learning, however, provide one method of narrowing the term's interpretation by convincingly pointing to a distinction between knowledge *about* something (declarative knowledge) and knowledge of *how to do* something (procedural knowledge; Ackerman, 1987).

To elaborate, *skills* encompass such diverse activities as skiing, typing, operating a machine, cooking, or doing a firm's financial consolidation and inventory management using a computer and special software. Ackerman (1988) argued that each of these skills is subsumed in a finite domain of behaviours. In addition, such skill specifications necessarily exclude a variety of non-motor-learned behaviours, such as genetic problem solving and abstract reasoning.

A working definition of *skills* must be established. Adams (1987) attempted to do so in a recent review of human motor skills research. His three defining characteristics of skill acquisition are as follows: "(1) Skills are a wide behavioral domain in which behaviours are assumed to be complex; (2) skills are gradually learned through training, and (3) attaining a goal is dependent upon motor behaviour and processes" (Adams, 1987, p. 42).

Can we assume then that skills are inherent in an individual? Sociological literature has suggested that skills are inherent in jobs instead (e.g., Hall, 1986, pp. 23-31; Kalleberg & Berg, 1987, p. 175). Sociologists have claimed that jobs have inherent skill requirements. Thus the primary mechanism through which "skilled" people select their work appears to be long-term occupational shaping rather than immediate or substantial changes in the structure of their work (e.g., Kohn et al., 1983; Spenner, 1983). Professional and tradespeople possess different skills and choose particular jobs that make use of those skills (Kohn et al., 1983). Furthermore, the comparison between professional and trade jobs suggests that there are major differences in the content and length of training – professional jobs emphasize theory while the trade jobs stress hands-on experience (Hall, 1986, p. 68).

According to Spenner (1983), skill is a multidimensional concept. Its two major dimensions are substantive complexity and autonomy. *Substantive complexity* refers to the level, scope, and integration of mental, interpersonal, and manipulative tasks in a job. Measures of these characteristics are available for thousands of jobs in publications like the U.S. *Dictionary of Occupational Titles* or the *Canadian Classification and Dictionary of Occupations* (CCDO). *Autonomy* refers to the amount of control the individual has over the performance of the job. Thus increased complexity requires more training for adequate performance; and the more difficult it is to supervise closely and routinize, the more autonomy over these activities must be given to the worker.

Developing a Conceptual Framework for Employee Skills in a Technology-Mediated Job

The above discussion indicates that psychological studies have mainly concentrated on the learning of skills (i.e., behaviours) through training (Adams, 1987), in which goal accomplishment is dependent upon the level of declarative and procedural knowledge needed to perform satisfactorily (Ackerman, 1988). Sociological thought shifts attention from the individual to the organization or job and states that the complexity and autonomy in one's job-related tasks define one's skills, without agreeing on a definition of the term beyond the general, as outlined here (e.g., Hall, 1986, pp. 23-26; Kalleberg & Berg, 1977, p. 176; Spenner, 1983).

To assess the interrelationship between technology-induced change and skill levels, it is necessary to develop a skills framework that includes the dimensions indispensable for analytical purposes, namely, the transferability of skills and the degree of complexity and abstractness inherent in job-related tasks. Hence, the complexity of the skill concept requires the merging of both psychological and sociological thought to increase our understanding. Figure 12.1 conceptually outlines these interrelationships between declarative and procedural knowledge on the one hand and job-related skill categories on the other. Each position requires all types of skills outlined, but the mixture of declarative and procedural knowledge for each type may differ. Furthermore, certain positions may require more depth (e.g., managers must know *more* about work planning than nonsupervisory personnel) and a wider range of skills within each category than others (e.g., managers need *various* conceptual skills to effectively plan work for themselves and others while a subordinate only plans his or her own work).

Figure 12.1 uses two dimensions to illustrate the job skills and job knowledge concept—namely, the level of transferability of skills and the degree of complexity and abstractness inherent in job-related tasks. Figure 12.1 lists, in descending order of transferability, the skill categories inherent in a job. In other words, task skills that are specific to one's work and cannot be transferred to another position as easily as basic skills come last. Moreover, as the level of complexity and abstractness increases, the amount of declarative knowledge needed to perform satisfactorily on the job increases.

Employee Skills and Technology Management 233

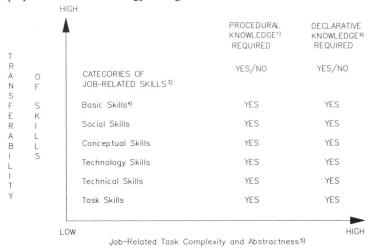

Figure 12.1. A Matrix Approach to Job Skills and Job Knowledge
1. Procedural knowledge is knowledge about how to do something (e.g., write an invoice or process a phone order).
2. Declarative knowledge could be defined as knowledge about something (e.g., theory and concepts). An increase in declarative knowledge increases the job complexity and abstractness of job-related tasks.
3. Managerial jobs may differ from support staff ones by having a greater range of skills within each skill category (e.g., a greater number of task and technical skills) and requiring a greater depth (a higher level of declarative and procedural knowledge) about the skill. Skills are listed in descending order of transferability from one position to the next.
4. These include reading, writing, and arithmetic skills (also known as the 3 Rs).
5. The greater the inconsistency between job-related tasks, the greater the task complexity and abstractness experienced by the employee on his or her job.

Basic skills are reading, writing, and arithmetic (the three Rs) generally acquired during formal education (primary and secondary schooling). *Social skills* include the ability of employees to organize their own efforts, individually or as part of a team, and, for managers, the ability to organize the efforts of their peers and subordinates as well as getting along within the work group. *Conceptual skills* include planning, assessing, decision making about task and people-related issues, and judging or assessing tasks done personally or by others (e.g., performance appraisal by supervisor and self — see Table 7.1; Ray, 1989).

Different competency levels are required to perform well in various positions. Therefore, the depth one requires in basic, social, and conceptual skills depends upon the type of work performed. The latter also determines how much declarative and procedural knowledge about a specific skill is required to perform well.

Technology skills encompass, but are not necessarily limited to, the individual using the technology appropriately to perform job-related tasks and knowing how to deal with, or preferably how to prevent, technology breakdowns. Furthermore, they include employees' correct handling of safety and legal issues and ethical considerations when using the technology.

Technical skills are an individual's physical ability to transform an object or item of information into something different. Again, it is quite likely that these skills are acquired during one's apprenticeship, during vocational training at college, or in some cases in the university (for professionals such as doctors and engineers). Some technical skills may also be acquired during firm-sponsored training (see Table 8.1).

Task[1] *skills* are required to perform job-related tasks and are usually specific. Thus transferability of task skills to another position or employer is often limited (e.g., knowing how to operate a certain machine for one company does not help in operating a different machine for another employer). Each position requires certain task skills that may have been previously acquired through formal education but are likely to have been refined during the socialization phase after joining the organization (see Table 7.1).

How does the above depart from earlier conceptualizations of skills presented in the literature? The approach outlined here makes an attempt to bridge psychological and sociological theory when developing a skills concept. The multidimensional approach as proposed by Spenner is, for instance, encompassed by the dimension in Figure 12.1 that measures the degree of complexity and abstractness inherent in a position (substantive complexity according to Spenner, 1983), and Ray's (1989) approach to the skill issue is expanded upon (e.g., Ray felt that managers needed technical, conceptual, and social skills). The advances of cognitive psychology are contemplated in this figure by including declarative and procedural knowledge (e.g., Ackerman, 1987 & 1988), which, in sociological terms, determines the degree of autonomy a person has in his or her work. A job with primarily procedural knowledge requirements (e.g., cashier) has less autonomy than one requiring substantial declarative knowledge (e.g., nurse and teacher).

The model outlined in Figure 12.1 raises several issues: How does the trend for specialization hinder the acquisition and maintenance of the job-related task skills? How may deskilling and upskilling be explained using the approach presented here? And, further, how may the structure of work processes influence skill levels? (See also Chapter 9.) These issues are discussed in detail below.

Structure of Work and Work Processes

Effective EUTM requires the firm to identify whether technology-induced organizational change may alter the overall work process and the structure of work for various types of job. Such changes could have a qualitative and quantitative impact, altering the skill requirements and task content inherent in jobs. In recent years, much attention has been paid to how new technology may lead to the phenomenon of "deskilling" of tradespeople (Braverman, 1974; Hall, 1986). Deskilling is seen as an attempt by management to assume control of work by depriving the employee of opportunities to use his or her skill, which leads to atrophy. Hence, integrated, self-controlled work by skilled tradespeople is replaced by centralized design, standardized procedures and products, and the fragmentation of skills into "detail" work in unskilled, specialized roles.

From this perspective, new technology would seem to change the range of skills needed by the employee for the worse, making work more fragmented and simplistic. Looking at Figure 12.1, this implies that the amount of procedural knowledge may increase while declarative knowledge decreases. Moreover, within a particular skill category (e.g., technology skills), the employee's mastery of various skills (e.g., programming and maintenance skills) decreases. Consequently, the individual may have a few technology skills (e.g., only knowing how to work with one or two software packages), with sufficient depth to perform technology-mediated job tasks but no more (e.g., the employee can use Lotus but does not know how to work Lotus graphics because it is not absolutely necessary to perform the job).

How does it all relate to technology-induced organizational change? A successful EUTM approach requires that the firm knows how much declarative and procedural knowledge its employees have about a particular skill category, as outlined in Figure 12.1. For instance, if the work force has limited declarative knowledge about writing and communicating (e.g., limited comprehension of grammar and rules about sentence structure, parallelism in sentences, paragraph construction, and conjugation), a new word-processing system will not improve the firm's written materials. Employees whose declarative knowledge about conjugation is so limited that they cannot put the correctly spelled unconjugated verb (as given by a spell-checker) into the required form to obtain correct sentence structure will not benefit from the most extensive, up-to-date dictionary available, even if they possess the procedural knowledge to run the program.

Job-related tasks and skill levels. As the previous example illustrates, different levels of *functional competence* in the various skill categories outlined in Figure 12.1 are needed to master certain jobs. *Consistency* of job-related tasks also affects the complexity of one's work. The greater the level of inconsistency between the various tasks being performed, the greater the complexity and abstractness of these job-related tasks and, most important, the greater the amount of declarative knowledge needed to comprehend the concepts and theories required to master the tasks.

The general trend affecting the distribution of jobs over the spectrum of skill and knowledge requirements is also of importance (see Sorge, 1989). In this context, what interests us most is whether the increasing use of manufacturing and office technology in the future will perpetuate employment subsystems, that is, work types such as (a) *knowledge worker* and (b) *skilled*, (c) *semiskilled*, and (d) *low-skilled employee*. And, furthermore, if it does, could a polarization occur, leading to a large percentage of workers holding low-skilled jobs while, for example, semiskilled jobs as a percentage of total employment decrease (e.g., Braverman, 1974)?

In the past, we have usually distinguished between industrial (blue collar), salaried (white collar), trade (skilled labour), and secondary employment subsystems (unskilled labour; Osterman, 1986) as well as professional jobs (e.g., lawyers and doctors; Hall, 1986, p. 68). The classification system suggested in the preceding paragraph, however, eliminates the distinction between blue- and white-collar workers. Instead, it distinguishes between skilled workers, including such occupations as potters, nurses and teachers, and semiskilled workers such as computer operators and mail carriers.

Of the four subsystems offered above, the knowledge worker category includes managers, R&D scientists, and engineers — professionals who acquire, process, and interpret information from a variety of sources to solve problems. This category also includes professionals who diagnose problems, such as doctors and lawyers. Thus a substantial amount of declarative knowledge is required, transferability of skills to another employer is assured, and the job's complexity level is relatively high. This work stresses understanding the concepts and theory behind certain skills as opposed to semi-skilled work, which emphasizes procedural knowledge. Nonetheless, consistency in one's work may range from high to low across jobs within this classification.

The fourfold classification system also differs from earlier efforts in this area (e.g., Osterman, 1986) because it tries to capture the effects caused by technology-induced changes in job content. We can see the benefits of this classification system by contrasting it with a traditional system. Computer software specialists design, write, and modify the instructions that make computers work. Using a traditional classification system, one would say that this is both a white-collar (Osterman, 1986) and a professional job (Hall, 1986, p. 68). This categorization is, however, inadequate to distinguish between different types of software positions. The degree of declarative knowledge required determines into which of the first three categories (knowledge, skilled, or semiskilled work) the job will fall. For instance, Kraft and Dubnoff (1986) found that software specialists doing maintenance/application work and system programming were semiskilled because their skills were firm-specific, job content was narrow (high component of procedural knowledge), and control was limited (low complexity level, high degree of routine in job-related tasks and limited transferability of skills). These researchers also reported that skilled software positions required designing applications based on specific customer needs and demanded frequent interaction with customers and peers (i.e., mixture of declarative and procedural knowledge). Software employees in the knowledge category (emphasis is on declarative knowledge) had to plan, manage and make decisions about software and hardware purchases and sales and had high complexity and transferability of skills. This example illustrates the usefulness of the four-category system outlined here — it can help in distinguishing between the various software positions, whereas the other system cannot.

Technology Management: ILMD and the Skill Debate

Although technology may lead to changes in the mix of declarative and procedural knowledge needed to perform a job, ILMD may put certain restrictions upon the firm's ability to induce changes without explicit consent by the workers (see Chapter 10). Moreover, technology acquisition does not have to lead to negative quality of work life outcomes for employees. Recent research tried to demonstrate that the implementation of computer-based technology in office settings is not necessarily driven by management trying to increase their control by narrowing tasks. Instead, managers are often quite concerned about office morale, because employee resistance towards end-user computing severely limits the benefits to be gained by computerization (Attewell, 1987a).

As the acquisition of new technology requires a substantial financial commitment, managers have to justify the investment by improving the organization's performance. To accomplish this while assuring quality of work life, managers must try to increase *relative surplus value* via new technology rather than trying to squeeze out *absolute surplus value* via speeding up the work process. Put differently, managers are trying to get their employees to work smarter to avoid having to work harder/faster (see Kling & Iacono, 1984). This approach requires employees to improve their depth in each of the skill categories outlined in Figure 12.1 to facilitate the achievement of desirable relative surplus values. Consequently, the employee may require additional declarative and/or procedural knowledge. Thus technology acquisition and organizational adaptation does not automatically lead to deskilling, as has been frequently assumed (e.g., Braverman, 1974).

In the United States, longitudinal studies have shown that job complexity affects intellectual flexibility positively, helping individuals to remain flexible in their thinking and encouraging the learning of new knowledge and skills, thereby aiding employees in coping effectively with change in their work and environment (Kohn et al., 1983, chaps. 3, 4). This suggests that highly specialized jobs that have limited job-related task complexity and abstractness affect intellectual flexibility negatively, thus curtailing employees' future potential for learning new skills and undergoing retraining successfully (Kohn et al., 1983). Consequently, a typist trained only as a word-processing operator may have some difficulty in acquiring the skills needed to make use of a fully networked computer system (e.g., accessing the mainframe to download or upload data; using various programs such as word processing, spreadsheeting, electronic mail, accounting, telecommunications, and graphics). Substantial declarative knowledge about the applicable skill categories may, however, help an employee comprehend work content and organizational changes caused by technology acquisition more easily. Accordingly, organizations need to encourage the achievement of declarative knowledge in several or all of the skill categories outlined in Figure 12.1 to assist future retraining prompted by technological developments. As a specific example, a firm in West Germany may have less difficulty in managing organizational adaptation of new technological innovations than its North American counterpart because German employees acquire more declarative knowledge during formal education and apprenticeship training than North Americans and thus may handle change more easily (see Bjorn-Andersen, Eason, & Robey, 1986; Casey, 1986).

Job design and the sociotechnical approach. From a quality of work life perspective, it is obvious that job design is an important dimension for a successful EUTM strategy. Technology-induced organizational change requires the firm to design jobs in such a way that they meet organizational and individual needs. One comprehensive framework for job design is the sociotechnical approach, which is based on the assumption that change affects work content and method as well as relationships among workers. Katz and Kahn (1978, pp. 716-717) distinguish between three discrete magnitudes: the

Employee Skills and Technology Management 239

job, the work group, and the organization as a whole. Organizational changes may affect all three magnitudes and be either social or technical. As an illustration, Volvo's change from assembly line to work-group production was a technical change, as the goal was to use new technology in the production process. When this technical change was implemented, social relationships within the group and between individuals and groups also changed.

EUTM can affect all three magnitudes, usually causing both technical and social change. Thus task, technical, technology, conceptual, social, and basic skills required to perform the necessary task(s) well are very likely to change. For instance, Attewell (1987b) reports that, after the introduction of interactive computing in the regional office of one of the largest U.S. medical insurers, the degree of coordination and interdependence increased. Workflow patterns changed, with paperwork going to an examiner, then to the data processors and entry clerks, then back to the examiner. This required increased coordination of the work flow but also resulted in certain tasks that had been part of the examiner's job being done by the clerks in the intermediate stage. The examiner's work shifted focus to encompass teaching clerks certain tasks; relationships between examiners also shifted, and a lot of informal advising began to occur.

Whereas some job design approaches look at individual attitudes, such as the job characteristics method (Hackman & Oldham, 1980), sociotechnical job design goes beyond individual attitudes. Such an approach is helpful because technology-induced organizational change can, and most often does, alter the social *and* technical aspects of work, not only by changing job roles but also by affecting work-group interaction and the organizational structure. To illustrate this further, if a customer service employee would like to have more variability and complexity in his or her job, the answer may not be so simple as enriching his or her job by increasing these elements. The current system might not allow accommodation of employees with various needs when it comes to variability and complexity. If the firm wants to design a work system that can accommodate this worker's needs, this will not only alter the work of the individual but also that of other groups as well as altering departmental structure (Hall, 1986, chap. 9). Because interaction between subordinates and supervisors differs depending on whether the subordinate employee has high or low levels of complexity in his or her job, the work group will be affected. Additionally, individuals with more complex jobs may also use different technology (hardware and software) to perform their duties. Hence, the most effective approach may be a sociotechnical job design that recognizes the changes in social and technical work dimensions, and especially considers individual, group, and organizational work dimensions affected when individual jobs change or technology is adapted (Hall, 1986, chaps. 3 & 9).

Culture, Market, and Employee Skills

As outlined in Chapter 10, both a country's and an organization's culture encompass symbolic vehicles such as meaning, beliefs, ritual practices, work forms, and ceremonies. Informal cultural practices such as "the grapevine," stories, and organizational rituals of daily work life are also included. According to Swidler (1986), these symbolic forms shape action strategies. Culturally shaped skills, habits, and styles explain what is distinctive about the behaviour of organizations, groups, and societies. Cultural stability characterizes an environment that sustains existing strategies of action; instability is inherent in change that causes new patterns of action to evolve.

How does organizational and national culture relate to EUTM? To begin with, a stable culture makes it more difficult to change the values and beliefs held in an organization. For instance, typing has usually carried some stigma and has been done mostly by women holding clerical positions (Morgall, 1983). One of the greatest obstacles for the success of computers in the office is their limited use by employees in upper hierarchic levels, who feel that it is beneath them to enter anything onto the computer. In an office environment with extensive computer use, an executive could type his or her report as a draft using the computer, make a printout, and, thereafter, make editorial changes on this hard copy. The secretary in turn would then make these subsequent changes on the computer file. Unfortunately, most executives have shied away from using their technology in such an effective manner and, instead, often write down their draft in longhand or else record their ideas with a dictaphone. Using this intermediate step, however, decreases the manager's effective use of available office technology.

The above illustrates that effective EUTM requires changes in habits and behaviours that are exhibited in most organizations and imbedded in the culture (e.g., managers dictate their correspondence to their secretaries who will type it). Increasingly, secretaries are no longer responsible for just one individual but, instead, may provide support to several people. Effective use of computer-based technology, therefore, becomes an important tool for the manager or professional to increase his or her productivity by doing certain tasks on his or her own computer. To promote computer use in the higher echelons, however, elimination of the stigma attached to typing is necessary.

Encouraging managers to use computers may also require some additional training, helping them to brush up on their typing skills (Gattiker, 1990c). Moreover, nearly all employees require some type of training to work with computer technology. In some countries, recurrent training is part of working life and most employees expect and welcome training updates during their working careers. For instance, 40% of the Swiss working population surveyed attended some type of continuous education in various subject areas (accounting, languages, computer skills, and so on) to increase their knowledge and skills, thereby improving their chances for continuous gainful employment (BIGA, 1988). These people upgrade their skills in various disciplines mostly on their own initiative. In such a cultural environment, it is easier for an organization to obtain a positive response from its work force for training opportunities in the technology area. In other countries, it might require some encouragement by the organization in the form of, for example, cash payments to employees attending training, to assure that employees sign up and attend a computer seminar offered or sponsored by the firm during or after working hours.

Training and its objective of improving employees' skill levels has been an issue for unions in numerous countries. For example, the *Schweizerischer Gewerkschaftsbund* (Association of Swiss Trade Unions, 1989) is demanding paid educational leave to encourage skills upgrading, and, in Sweden, this has already been realized by trade unions for all Swedish workers (Rubenson, 1983) and to some lesser extent this is also the case for Dutch employees (Emmerij, 1983).

Industry, market and job skills. Up to this point, no definite conclusion has been reached about technology's effect upon the skill levels exhibited and job specialization experienced by the work force. The type of industry and the market may, however, help explain EUTM effects somewhat better (Attewell, 1987a). For instance, batch production requires more intellectual flexibility than mass production (e.g., assembly line work). Zicklin (1987) reported that using computer-numeric control (CNC) machines in the production process could lead to either deskilling or upskilling depending upon the job holder, corporate culture, and type of production used. In most cases, batch production led to upskilling when CNC techniques were introduced, because employees were more often required to make decisions and be responsible for the machine.

Maybe the most important factor is the market in which the firm sells its products. For instance, a *newly emerging market* for a product, such as telephone answering services, is often characterized by immature or dynamic process technology. The organization makes rapid and radical innovations and experiences growth. A more "stable" organizational culture would make acquisition of new technology more difficult because, as outlined earlier, it would decrease acceptance of the inevitable change in social and technical dimensions of work. If a complex work environment (job complexity and interaction with various groups and customers) exists, only the intellectual flexibility of employees can help them cope with such an environment and allow a technology-based work environment to develop and advance (Kohn et al., 1983). Increased intellectual flexibility in the employee facilitates the comprehension and acceptance of more abstractness—highly integrated technology makes it quite difficult for an employee to physically see how the evolutionary stages of the production process lead to the final product. Employee tolerance towards such abstractness is one crucial ingredient for effective EUTM.

In an *established market*, competition is quite intense. Accomplishing differentiation of products and services through innovation and technological changes enables the firm to remain competitive. North America's financial industry may fit this category. Banks and other institutions try to differentiate themselves from their competitors with unique products and better customer service. Consequently, a reduction in operating costs becomes necessary to attain acceptable profit levels as customer service costs have risen. Certain processes will, therefore, be mechanized (bank teller machines) while others may be automated (e.g., customers using the computer to access checking accounts, making automatic payroll deduction for the monthly rent/mortgage payment).

Walsh (1989) reported that in a mature market with standardized products, deskilling may occur. However, workers who have some control over the introduction of technology may, through ILMD, reduce any potentially negative effects of technology upon their skill levels.

In a *mature market*, price competition is accomplished via *cost-cutting* (materials and personnel) and product differentiation (e.g., Fiat versus Volvo). The emphasis is on speed in the production process, and the process technology is highly developed. Routine clerical costs constitute a large percentage of total expenses, and mass production helps decrease costs and speed up production. In such a situation, negative outcomes can result, causing some adamant resistance to the technology (Attewell, 1987a). One example is today's mail-order business, which has matured considerably in the last decade in most industrialized countries, becoming cutthroat. Data-entry operators must process a certain number of orders per hour to keep operation/processing costs per order at an acceptable level. These workers often complain of back pain and eye strain and quality of work life is extremely low (see Stellman, Klitzman, Gordon, & Snow, 1987).

When production becomes fairly standardized with the help of technology, the mastery level of procedural knowledge required to perform well on the job increases, and the range of skills held within each skill category is narrowed. Nonetheless, firms may differ, as the example comparing Fiat and Volvo shows (see also Chapter 10). Although sequential work flow at Fiat may reduce the range of skills, Volvo's use of parallel group production requires that employees attain at least semiskilled, if not skilled, job status. These skill levels are required for employees to work in a group in which tasks are rotated between members and where budget, quality control and maintenance duties must also be managed by the group (Van Houten, 1987). As outlined earlier, ILMD and national differences (e.g., labour supply, laws for participation by employee representatives in the decision-making process) explain, in part, why two firms in the same market may use technology either to increase or to decrease workers' skills (Walsh, 1989).

Technology management culture. Cultural differences may help further explain different technology effects for two organizations, even though the market scenario may be the same for both. The microcomputer became a real force around 1979 when Apple launched its Apple II. Between 1981 and 1990, changes in the type of microcomputers available to end-users have been tremendous. Companies who started to install microcomputers in 1979 have a different technology management culture today than a firm that started only a few years ago. People who have been working with a computer for a decade or more have been getting used to the incessant changes occurring in hardware and software. Since they started working with computers, some employees may by now have moved to their second or third machine, replacing "outdated" versions. Moreover, these workers today are probably utilizing the sixth upgrade of one of their frequently used software packages. Improvement in hardware has opened new possibilities for software and applications. Today's lap-top unit with a colour monitor, running on a 32-bit chip and a hard disk, with a battery lasting easily for four hours and a weight of about four kilograms or less, is bringing the computer into all kinds of professions and work environments. Employees can even take the machine with them on the road. In contrast to a computer novice, experienced computer users are more likely to accept such a technology and, more importantly, know how it can help expedite their work (Carter, 1990; Eriksson, 1990).

It is obvious that a more *mature technology culture* will evolve over the years with consistent use of a particular technology (e.g., computer-aided manufacturing and/or office technology). Values, beliefs, and rituals in regard to technology-mediated work and technology-induced organizational change will thus be understood by the majority of employees. In an *emerging* or *changing technology culture*, however, individuals may not have the cultural action strategies to facilitate personal adjustment to organizational change.

The previous description implies that differences in a technology culture may also evolve out of the development stage of an economy. For instance, advanced technology may not be readily available in a less developed country (LDC) because labour costs are low or import duties are astronomical, thereby making technology a high-expense alternative. Or, if it does possess the technology, an organization in a less developed country may have just begun using computers (see Rodrigues, 1988), so its employees are just starting to acquire the skills and tools needed to work with the technology. A lap-top computer may, therefore, be of little use in such an environment, while in a European setting, these machines become part of a person's working tool kit (e.g., sales representatives and service technicians and in police patrol cars).

Rodrigues (1988) did a study comparing Brazilian bank employees and supermarket workers' perceptions of their skill levels following the introduction of computer-based technology at their workplaces. In Brazil, both types of businesses exist in established markets. The results indicate that bank employees felt skill requirements became higher after the introduction of computer-based technology, but supermarket workers felt deskilling had occurred. This indicates that technology cultures may, first, differ based on degree of maturity achieved in using a particular technology (length of time and personal experience). Second, the type of industry and market may also affect the firm's technology culture. Specialization, upskilling, or deskilling can, therefore, be quite different when comparing two banks (same industry and market) or a bank and a supermarket chain (different industry and both firms in an established market). Unfortunately, research to date has tried to gain new insights into technology by making conclusions from studies using different types of organizations. For example, Turner discusses previous research in the computer-based technology area and, based on comparisons of results, summarises various findings (Turner, 1980, p. 31). He does not, however, mention that the differences reported may in large part be due to comparing firms in different industries (e.g., manufacturing versus service) and markets (evolutionary versus mature). The third distinguishing factor is the economy itself — a distinction must be made between an industrialized and a less developed country. The fourth factor is cross-national differences based on cultural factors, which are discussed below.

Cross-national culture and technology management. Technology's effect upon job content and skills may also differ based on national factors (see Chapter 10). However, cross-national culture has generally been ignored in discussions of research results. For instance, Bjorn-Andersen, Eason, and Robey (1986) report on a large cross-national research project. The countries (and type of company) are the United States (airline), Austria (bank), Denmark (electronics firm), and the United Kingdom (hospital). The authors attributed differences between firms surveyed to various information technology environments. Differences cannot be attributed to type of industry or technology alone, however, because cultural variation between the participating countries may be responsible for most of the disparities reported.[2]

Another difficulty is that many conceptual frameworks addressing the technology and skill issue evolve with a single country in mind. For example, Piore and Sabel (1984) base their argument — that work restructuring and the use of programmable technology should increase the importance of shop-floor skills — on U.S. data. They offer three reasons for their argument. First, workers should play a critical role in debugging programs or intervening when production goes awry; second, process and product innovation requires skilled worker knowledge of production; and, third, workers with broad skills are necessary to help master new responsibility caused by ever-changing product lines. Similarly, Sorge (1989) developed a contingency and workflow model to try to explain skill changes based on research conducted in France, West Germany and Great Britain.

As the examples below will show, however, Piore and Sabel's work may not necessarily be helpful in explaining skill and knowledge changes outside the United States, and Sorge's model may not apply beyond Europe. Until the validity of these models has been established beyond the one country studied, their generalizability is questionable (Kohn, 1989).

Research in the United States has sometimes contradicted Piore and Sabel (1984) but in other cases studies have provided support. For instance, Shaiken, Herzenberg, and Kuhn (1986) observed a U.S. sample and found that even in small batch production, there was little sign of the new "craft worker" described by Piore and Sabel. Furthermore, a premium placed on quality and fast delivery would require using technology and shop-floor skills in a complementary fashion. These authors did not find such a combination; instead, programming changes were not made on the shop floor but by a separate programming department. Kraft and Dubnoff (1986) discovered a similar fragmentation when studying U.S. computer software workers at the leading edge of the computer revolution. Again, they uncovered clear specializations that actually support the "deskilling" argument (Braverman, 1974). In contrast, when investigating the insurance industry, Attewell (1987b) reported that complexity for office clerks increased with new technology. In a study of meat cutters' changes in job-related task skills over a 40-year period, Walsh (1989) found that, in structurally diverse markets (e.g., retail industry), standardized products could not be developed, therefore, technology-induced work-flow changes increased meat-cutters' skills.

Okubayashi (1986) reported that, in Japanese firms, job content in manufacturing differed based on the phase of technology use (time since it was adopted and installed). For instance, operators of computer-aided production machinery had less complex jobs during the technology introduction period (responsible for operation and arrangement only), but during the starting period duties increased (also responsible for necessary troubleshooting) to reach their highest level during the stable period (additionally performing the necessary maintenance work on technology-based equipment). Furthermore, Japanese firms assume that operators of computer-based machinery must be retrained every two years to assure satisfactory skill and knowledge levels, as outlined in Figure 12.1 (Okubayashi, 1987).

Using a British sample of companies, Campbell and Warner (1988) reported that technology effects upon skill levels and training may differ across industries. Most important, they suggested that a core of industrial workers may have the benefit of increased training and skill levels but their "peripheral" counterparts will be left out. Based on their work and our earlier discussions (Chapter 6), it is obvious that part-time and term employment workers will most probably belong to the peripheral group of employees not receiving technology-related training, thereby lowering their potential for gainful employment in the long term (i.e., becoming "losers" as defined by Roskies, Liker, & Roitman, 1988).

Although increased use of technology does affect skills in many countries, the effects differ across nations. Moreover, within countries, differences may vary based on industry. Additionally, identical advanced technology will be used differently in the People's Republic of China than in more industrialized countries like the United Kingdom (e.g., Warner, 1987). Because technology affects organizational and management structures (Sorge, 1989), the complexity and abstractness of one's work and the transferability of skills are influenced as well. Furthermore, competency levels required for the skill categories outlined in Figure 12.1 and the appropriate mix between proce-

Employee Skills and Technology Management 247

dural and declarative knowledge varies across countries, markets, and industries as well as firms.

Another important factor for describing and categorizing EUTM developments is organizational type. Depending upon the firm's quadrant position (see also Tables 3.1, 4.1, and 4.2), it will introduce technology differently and outcomes may differ (e.g., depth and range of skills, levels of declarative and procedural knowledge).

Developing a Scheme for Technology Management and Employee Skills

Do the four different types of organizations in the quadrants presented in Figure 3.1 manage technology differently? How are these firms managing skill development in their work force in the more general and more specific technology domains to improve their competitiveness? How is the achievement of desirable skill levels by employees facilitated or restrained by ILMD levels? Will level of choice in SHRM influence EUTM and skill levels of a firm's work force? These matters are addressed in Table 12.1 and discussed below.

Organizational types as outlined may, however, not apply perfectly in each case. Moreover, cross-national (cultural) factors may further result in slight differences not captured in Table 12.1. Nevertheless, the table should help us structure the discussion and thereby facilitate our understanding of the complex interrelationships between organizational type, production processes (structural or parallel), and necessary skill levels for the work force to attain effective EUTM.

Quadrant A

This type of organization, as described earlier, is often bureaucratic (Table 6.1) and its culture quite stable. The firms tend to be publically owned. An example might be a government agency (e.g., health services) or, in Europe, one of the many state-owned railway companies. The market is mature and competition is stiff (e.g., railway cargo against private trucking firms). ILMD may result in a remuneration system that pays according to the skills inherent in a job. Hence, deskilling with the help of technology would be advantageous, reducing wage costs and increasing the firm's competitiveness. Most probably, however, high ILMD will prevent the organization from

TABLE 12.1: Four Human Resource Management Strategy and Internal Labour Market Types: End-User Knowledge and Skills

Variable	Quadrant A *High ILMD Determinism, Low Choice in Strategic Human Resource Management*	Quadrant B *High ILM Determinism, High Choice in Strategic Human Resource Management*	Quadrant C *Low ILM Determinism, High Choice in Strategic Human Resource Management*	Quadrant D *Low ILM Determinism, Low Choice in Strategic Human Resource Management*
Work types and technology-induced changes	Distribution within work force remains relatively stable due to high ILMD	All work types are present; deskilling, upskilling can occur	Knowledge employees and skilled work are likely to increase	Low and semiskilled work will increase
Strategy for end-user technology management	Employee focused	Employee market focused	Fully comprehensive—most stakeholders' concerns are an integral part of EUTM strategy	Market or "government" focused
Knowledge —Declarative	Changes at employee's initiative but difficult because job reclassification may be necessary	Will increase mostly due to ILMD pressure (e.g., union demands); philosophy of lifelong learning	Demands will be hightened for new job applicants: Tenured employees are supported by firm in their educational efforts to improve declarative knowledge; philosophy of lifelong learning	Firm-initiated efforts are sparce; firm support for employee-initiated efforts to acquire additional declarative knowledge sparce

—Procedural	Is changing to a limited extent because jobs must first be reclassified	Will increase/change only with workers' (union's) consent; philosophy of lifelong learning	Acquired primarily on the job; appraisal system provides only limited data to determine an employee's weaknesses; efforts seem quite unfocused
		Company-sponsored training given, used as screening device for new employees; philosophy of lifelong learning	
Skill categories —basic skills —social skills —conceptual skills —technology skills —technical skills —task skills	Skill levels remain stable although deskilling could occur for standardized goods	Skills upgrading possibly accomplished through job enlargement and/or job enrichment	No clear strategy, differences may occur between divisions
		Job enlargement and enrichment results in skills upgrading	

lowering the pay of current job holders, making reclassification an option only for newly hired workers.

As outlined in Table 12.1, the firm's EUTM strategy is employee-focused, because the high ILMD assures employee input before technology changes can occur. Also due to ILMD, the potential for an increase in low-skill jobs is limited. Nonetheless, reclassification of jobs may be possible if a two-tier wage system is established (see also Chapters 8 and 9), guaranteeing tenured employees the skilled classification (e.g., technician) and higher remuneration while new employees are hired at the low-skilled job level (operator).

An employee with initiative could change his or her level of declarative knowledge (e.g., additional formal education such as university courses). However, this could trigger demands for higher pay based on knowledge and skill levels, so it is unlikely that an organization would encourage it. The same reasoning explains the limited effect of EUTM on procedural knowledge; the firm tries to avoid extra expense.

Bureaucratic firms with cumbersome internal structures and processes (see Table 4.2) result in employee competency levels in different skill categories that are just adequate for satisfactory performance. Workers have a narrow skills range but some depth in their areas of specialty, thereby possibly limiting their intellectual flexibility and willingness to acquire new skills. Using the European railway example, this scenario of narrow job classifications (high ILMD) may cause additional deskilling, because intense competition requires the railway to reduce costs and increase the speed and accuracy of its cargo services. Hence, if permitted by employees and unions, the computer is used to automate tasks such as cargo management and train car constellation (location of cars on the train based on weight, type of cargo, and final destination). Manpower may be saved, and/or reclassification of jobs (lower skill requirements than before computerization) may allow for hiring new employees at lower wages.

Quadrant B

For a Quadrant B firm, the EUTM strategy is market and employee focused. Moreover, depending upon the type of market the firm is in (newly emerging, established, or mature), EUTM efforts differ. In addition, high ILMD requires the firm to respond to or even anticipate employee concerns early to reduce potential resistance. Having high choice in SHRM allows the firm to reclassify certain jobs, thereby permitting it to hire semiskilled employees at a lower wage. Nonetheless, high ILMD assures that job classifications remain stable, leaving the distribution across skill levels (knowledge, skilled, and semiskilled employees) relatively constant.

Knowledge levels will increase primarily through union demands. Furthermore, if the distribution between procedural and declarative knowledge has to change due to technology, the company will have to inform employees about this change. Workers will be informed of, and helped to understand, the need to foster the change process.

A Quadrant B firm tries quite hard and probably with success to have its work force achieve a skill competency level that will give it the range and depth needed to remain competitive and facilitate potential technology-induced organizational change. To make use of advanced computer-based technology and to help improve customer service, employees should show some innovativeness in technology use and take over end-user responsibility for most technology applications in this organizational type. Offering employees upgrading and enrichment may increase perceived quality of work life. In turn, this should increase employees' level of technology acceptance. The organization does, however, experience high ILMD, which could mean that some employees may resist the upgrading and enrichment of their jobs. In particular, the full-scale sociotechnical approach to job design may scare some employees at first, because the added complexity and responsibility may not always be appreciated. Nevertheless, introducing technology with a human focus will help this firm undergo a smoother changeover to advanced technology.

Earlier, we used Sony to illustrate this quadrant. Its high ILMD is based on a company-based union and lifelong employment offered to its workers. Nonetheless, the EUTM takes employee and market contingencies into careful consideration and the firm continuously helps its employees keep their skills up to date, thereby facilitating the effective use of new technology.

Quadrant C

Some Quadrant C firms may be found in emerging markets, but most are in established ones. Although established industries (e.g., steel and chemicals) as a rule tend to have to cope with strong ILMD, firms in this quadrant enjoy atypically low ILMD. A developing market and/or industry (e.g., semiconductors) requires continuous change and substantial innovation on the organization's part, and lower levels of ILMD facilitate this process. Entrepreneurial-type firms are most likely to be located in this quadrant. Based on the above, it seems understandable that a fully comprehensive EUTM strategy will be used, which takes into consideration the important stakeholders' interests and needs while trying to plan technology acquisition and its later integration into a firm's production/transformation processes.

As outlined in Chapter 6, firms in this quadrant are in the fortunate position of being able to discriminate in the recruiting and selection process and reward chosen applicants well. Hence, these firms have little difficulty attracting enough qualified applicants to fill open positions. Also, being in an emerging or established market forces the organization to adapt frequently and provide better service than its competitors. Because the firm's human resources provide it with the necessary competitive edge, knowledge worker and skilled employee jobs will become more numerous with the increased use of advanced technology.

The low ILMD facilitates technology-induced job changes that require new or additional declarative and procedural knowledge. Depending upon the firm's market power, it will use newly required knowledge for performing job-related tasks as a discriminating device for new job applicants (see Windolf, 1986). Moreover, continuous educational efforts are encouraged and supported, thereby instilling the concept of lifelong learning in all employees.

Generally an organization in this scenario will try to promote the accomplishment of higher general skills, especially technology skills, by upgrading positions and providing job enrichment and enlargement. Nonetheless, technical concerns may drive EUTM, thereby inviting resistance by the work force. In other words, the human focus might shift to the background and technical possibilities and solutions may be allowed to determine the EUTM strategy.

Credit Suisse has already been mentioned as an example for this quadrant. As a traditional and relatively old (over 100 years) bank, it still enjoys low ILMD. Its employees have substantial technology skills (i.e., information technology, both technical and software know-how) and the bank is trying to increase use of technology, especially in computerized customer services (e.g., computer-facilitated interaction between the customer and his or her accounts, processing orders such as transfers and payouts automatically without any human labour by the bank). The firm's competitive strategy is based on a philosophy of lifelong learning and constructive use of technology, with an EUTM focus that tries to balance customer and employee needs as well as ethical considerations in using the technology.

Quadrant D

As outlined in Chapter 2, the organization in this quadrant may not yet have developed its strengths and found its niche for peak performance. Hence, the firm is probably not a forerunner of new technology. The company's lack of a clear EUTM strategy may leave it floundering. In addition, the introduction of various technologies in different departments without clear coordination to assure compatibility (e.g., computer-based technology allows up-loading and down-loading between different host computers; as a result, computers can "talk" to each other) can occur and hamper EUTM effectiveness.

As outlined in Table 5.1, there is always the chance that low ILMD experienced by firms located in this quadrant could lead to some degree of exploitation of human resources. Furthermore, the market is likely to be mature, and standardized products are the norm. Consequently, low-skilled and semiskilled work will increase.

Educational activities initiated by the firm or by employees to increase or change procedural and declarative knowledge are rather scarce for several reasons. Support of employee-initiated educational efforts is limited, and the lack of a clear job classification scheme makes it difficult for the firm to determine the appropriate knowledge levels required for a competent performance. The appraisal system, as discussed in Chapter 7 (see also Table 7.1), is incapable of providing the necessary data for such decisions.

Job enlargement may occur because the firm could perceive such an approach as a simple way to assure better labour productivity. Nonetheless, the lack of a precise strategy will make the firm vulnerable. For this quadrant, we used the Swiss PTT example earlier. PTT has never been a technology forerunner and its workers do not have clear ideas about how computer-based technology such as fax, letter and package sorting, and telecommunication will affect their jobs in the future. However, further deregulation of the postal and telecommunication monopoly seems imminent (e.g., private parcel delivery services, a variety of equipment suppliers, and subleasing of telephone lines), making acceptance of current technology levels imperative. Future developments should prove interesting in various ways, and no strong union has yet emerged to make itself the champion of the resisting workers' cause.

Summary

This chapter has outlined and illustrated why management of technology and employee skill development may be some of the biggest challenges faced by firms in their efforts to remain competitive. Relating the issues to previous chapters shows that organizational type influences the firm's possibilities for a comprehensive strategy for technology management. Table 9.1 suggests that the organizational remuneration and reward systems used by Quadrant

B and C firms are more likely to simplify technology acquisition and organizational adaptation. Additionally, these firms put a higher emphasis on recurrent training for their work forces (Table 8.1), thereby continuously assuring sufficient skill breadth to cope with technological changes in the workplace. Selection systems in these quadrants may already attract the type of employees who thrive on this kind of company environment.

Although frequent technology-induced organizational change requires continuous upgrading and individual willingness to learn and face increasing or decreasing job complexity (e.g., Kohn et al., 1983; Spenner, 1983; Zicklin, 1987), a firm's technology strategy must fit the skill competency and knowledge levels attained by its work force. Thus early opportunities for adopting advanced technology are not effective if the work force does not have the necessary skills and tools to use the technology. Organizational types A and D have a distinct disadvantage compared with types B and C. Technology acquisition may by limited by ILMD in type A firms, hindering continuous technology acquisition, organizational adaptation, and development of the firm's strength in the technology domain. Certain EUTM strategies may not be easy to execute because ILMD may provide employees with the means to prevent the organization from going ahead with technology acquisition and organizational changes resisted by the work force. Type D firms are probably the more vulnerable of the two. No clear direction in EUTM strategy is apparent and support systems, such as training and career development (see Table 9.1), may not have been put into place yet. Thus falling farther behind home or internationally based competition in EUTM is practically guaranteed, unless the firm can move to another quadrant.

Another important issue is that firms in Quadrants A and D are often government owned or controlled (e.g., Quadrant A—Germany's DP, Quadrant D—Swiss PTT; see also Table 5.1). Their market position is usually dominant and the market is mature which can have some negative effects upon skills. For instance, the DP's annual technology investments are as high as the six largest West German firms combined (Möschel, 1989). However, since the DP has succeeded in developing a market for standardized goods (e.g., postal and telecommunication services), the likelihood of deskilling effects upon its employees is enormous. Space limitations prevent us from further discussing this issue (please refer to de Pay, 1989; Williamson, 1985, and any macroeconomics textbook for further information on this phenomenon).

It is obvious that the literature has looked at technology management but, unfortunately, has not considered cultural as well as social comparison processes. For instance, research in technology management has not identified or accounted for culture-, industry-, and market-related variables when

Employee Skills and Technology Management

comparing firms within and between nations. Thus interpreting results is difficult because some of the significant effects reported may be due solely to these intervening variables. Moreover, environmental factors (e.g., as proposed by contingency theory), ILMD, and choice in SHRM are not usually mentioned in such work. Technology management theory and research have ignored most of the sociological, psychological, and behaviourial attempts to deal with EUTM in organizations. The result has been a flood of research with a limited foundation in theory and in the work of other disciplines, with few notable exceptions (e.g., Long, 1987; Sorge & Warner, 1986).

How does this chapter relate to Figure 2.1? Figure 2.1 suggests that skill levels depend in part upon the subjective and objective culture. Furthermore, environment and choice experienced by the firm influence the organization's EUTM strategy and the skill level (technology and other skills) attained by its work force. Looking back, employee skills and technology management relate in numerous ways to the previous parts of this book. Looking forward, Chapter 13 develops a conceptual framework for studying technology acquisition and organizational change by making the individual the central focus.

Notes

1. Because of the ambiguities of the word *job*, it seems preferable to use the word *task* instead. *Job* may refer to the activity of performing the task, the task itself, or the employee's occupation. Even though *task* contains its own ambiguities, it seems a more precise word.

2. For instance, in the United States, the hospice industry (which provides nonacute care for the elderly and terminally ill) is part of an emerging market, but hospitals (private and others) are in an established market (differences between industries) — yet both types of organizations are usually located in Quadrant A. The U.S. airline industry (mostly Quadrant C, with low determinism and high strategic choice) has been riding a roller coaster in terms of market type. After becoming almost complacent in its mature, established market (about 30 years old), deregulation in 1980 gave the industry a newly emerging market situation again. It developed into an established market by the mid-1980s, and has now evolved once again into a mature market dominated by large carriers. In contrast, free trade between Common Market countries in 1992 will probably bring some deregulation, which may result in the member countries' airline industry becoming an emerging market and possibly evolving into a mature market, much like what happened in the United States. Hence, similar industries may be in different markets when cross-national comparisons are made.

13. Development of a Theory

There has recently been a great deal of reflection and introspection concerning the state of theory when it comes to technology acquisition and technology-induced organizational change. Literature has focused on the direction the field will take in the coming decade (e.g., Ettlie, 1988), and authors have paid attention to explicating the various technology paradigms used in research as well as their methodological linkages (Doswell, 1983; Panko, 1987). Research has also commented on past efforts to explain the innovation process and technology acquisition by various firms (e.g., Gattiker & Larwood, 1988; Urabe, Child, & Kagono, 1988). These efforts are especially striking considering that this type of self-conscious examination of the state of technology management and technology innovation theory in print has been lacking in the past. For real progress to be made, however, knowledge and paradigms used in various other disciplines must be integrated to help facilitate the understanding of technology acquisition and organizational adaptation. Particularly, further work is needed to conceptualize these metaphors and paradigms into a theoretical framework and apply them to the technology management domain in the office setting.

This chapter offers a strategic set of approaches to analyzing technology acquisition that is consistent with the relational, physical, and technological realities of organizations that must manage technology acquisition and organizational adaptation. These perspectives centre around office work and the office itself, structure and processes, effectiveness, and also technology acquisition. They are examples of the kinds of analytical approach that, along with the theories and other approaches reviewed earlier, hold promise for advancing our understanding of technology acquisition in the office setting with specific attention to technology management. This promise is largely unfulfilled because, for the most part, we have conceived of technology acquisition as being managed by rational executives and employees. I hope that the previous sections of this book have, however, illustrated that such an approach, when put to practical use, often fails to lead to readily comprehensible explanations of technology acquisition and organizational adaptation.

Up to this chapter, the book has dealt with technology acquisition in organizations for any work setting (e.g., office or factory floor), but now the focus is primarily on office work and technology-induced organizational change. In industrialized countries, office work has become, or is in some cases on its way to becoming, the most common type of work for employees. An office job is made up of numerous tasks inherent in a particular function being part of the overall organizational structure. As outlined in Chapter 12, skill levels within an office are different depending upon the type of tasks to be performed. In the past, office work was characterized by a desk, office chair, paper and pencils, and some limited technologies (e.g., calculator, typewriter) that were needed to perform one's job in a particular physical location within a building. This has changed; and to better understand technology acquisition and end-user technology management (EUTM), our conceptual framework about office work must be improved to include it. Even though this chapter predominantly concentrates on the office domain, the major principles discussed below apply for any work setting (e.g., factory floor or retail store setting).

Figure 2.1 illustrated that the approach taken in this book to EUTM is on both objective-subjective and macro-micro dimensions, thereby permitting the analytical framework to be more comprehensive than otherwise possible. Accordingly, we must develop a theory that puts the individual at the core and tries to develop a structure that encompasses the thinking of various disciplines to increase our understanding of the phenomenon to be investigated, namely, technology acquisition and organizational adaptation.

Office Work and Technology Management

The first compelling empirical reality for organizations is that they are physical entities that consist of manufacturing or production facilities and offices. In the past, research has mostly concentrated on the factory and not the office. One reason may be that, historically, the office has symbolized secondary activity, with production playing the dominant role in manufacturing (Sydow, 1985, p. 24). However, the growth of the service sector (e.g., banking, finance, and insurance) has resulted in office work becoming a primary activity for many organizations. Technology acquisition has led to rapid change and reorganization of such offices and their processes.

There is not yet a clear definition of the *office*. In fact, some have claimed that our knowledge about office processes may still be medieval (Doswell,

1983, chap. 1), making any advanced conceptual framework or the development of an all-encompassing research paradigm an impossibility.

One difficulty involved in developing a concept for EUTM in office settings is the existence of various paradigms in the field. According to Kuhn (1970), the term *paradigm* refers to the values, beliefs, and techniques shared by members of a scientific community. Empirical studies have reported differences in activities and attitudes of researchers working in fields with higher as opposed to lower levels of paradigm development (Pfeffer, Salancik, & Leblebici, 1976). In general, members of fields with higher levels of paradigm development (e.g., physics) showed attitudes and activities reflecting a great deal of consensus over theories, research goals, methodologies, and curricula. However, in fields with lower levels of paradigm development (e.g., organizational sociology), beliefs about these issues are rarely shared and conflict prevails over consensus (see Hickson, 1988). Thus well-developed paradigms function much like ideologies, binding people together, helping them understand their worlds, and providing them with guidelines for appropriate behaviour.

The domain of technology research suggests that paradigm development in the discipline is in its infancy. Most of the literature has been struggling to find some common ground. However, paradigm development is an important ingredient for advancement in any discipline. More consensus over the concept of the office and technology in particular, as well as over the methodologies needed for such research, is essential.

Without common ground, it becomes increasingly difficult to achieve a proper conceptualization of those elements of a theory that are crucial in the choice of a proper comparative strategy of verification. For instance, North America has seen a heightened interest in the high-technology industry and, therefore, research abounds (Kleingartner & Anderson, 1987; Link & Tassey, 1987). Variables used for defining such organizations include, but are not limited to, the following: (a) the firm must have a higher proportion of knowledge workers in engineering than found in other manufacturing firms; (b) production processes are based on applications of science; (c) R&D is important for success; and (d) markets for the firm's products must be both national and international. It is not clear what part of the concept of high technology the above variables are supposed to measure; neither can operationalization of these variables be accomplished without difficulty. For instance, what percentage of engineers makes a company part of the high-technology group? In addition, it is hard to accept that high-technology firms exist in the manufacturing sector only; genetics and medical science firms may also use high technology but do not belong to the manufacturing sector.

Development of a Theory

Belous (1987) came up with a definition classifying high-technology firms as having an R&D to sales ratio at least twice as high as the overall industry average and a rate of employment of technology-oriented workers three times as high as the industry average. This definition includes companies in computer-, drug-, chemical-, and aircraft-manufacturing-related areas. Nonetheless, according to Belous (1987), in the 1990s only 5.3% of the total work force in the United States is projected to be part of high-technology firms.

The difficulty with developing a paradigm for technology management is also reflected in the business community itself. For example, a study done in Europe reported that 22% of small- and medium-sized firms surveyed felt that they were high-technology companies; 53% felt that they were middle-technology firms; and only 17% assessed themselves as low technology (Pfister, 1988). This assessment is characteristic of the status given to high technology, although many high-technology firms have had rather dismal success recently (e.g., Tecan and Wild-Leitz of Switzerland and Nixdorf Computer AG in West Germany reported losses for 1988 and in 1989 the profit picture did not improve very much). The failure of some European high-technology firms to succeed cost their shareholders dearly (e.g., by mid-1989, some shares had dropped by more than half since their before-crash high of October 1987).

Neither researchers nor public opinion have come up with a clear definition of a high-technology firm. Instead, public opinion and business show that it is a sign of the times to be high technology focused even though, in some cases, low-technology firms are doing quite nicely. Government response in various countries has been to increase funds for basic and applied "high-technology" research (e.g., Canada, France, and Switzerland). Because paradigm development has fallen by the wayside, numerous definitions of high technology and EUTM exist without reflecting much agreement over theories and concepts. Moreover, the empirical body of literature lacks the common ground that would allow a comparative strategy of verification.

These difficulties suggest that high technology may be implemented as a concept more easily by looking at the type of products generated with high technology — such as machines, knowledge (e.g., design and software), and organisms (e.g., genetic engineering). Furthermore, production could occur in any of three settings — factory, laboratory, or office. Thus once the above parameters have been defined thoroughly, a comprehensive definition of *high technology* can be provided to tremendously facilitate empirical work in this area. Currently, empirical data is of limited use as nearly every study uses a different definition for *high technology*.

Technology and the Office

The above discussion suggests that the term *high technology* is a misnomer and paradigm development should concentrate on finding common ground. The rudimentary framework given above is the first step. It will facilitate the achievement of a proper conceptualization of those elements of a theory of technology management that are crucial in the choice of a proper comparative strategy of verification.

Reviewing the literature dealing with office technology, information systems, and contingency theory, it is clear that the literature makes a distinction between the analytical and the interpretive views of offices. Although the former is a more task-oriented concept, that is, office employees perform certain tasks or take on certain roles, the interpretivist views are more socially focused, trying to increase our understanding and knowledge about offices by, for example, looking at office work within the context of behavioural and social interaction (Hirschheim, 1985, chap. 3).

Others have suggested that offices are nothing more than an organizational subsystem that processes input to produce output. Hence, they are open systems assumed to have the processing of information as a central characteristic (Long, 1987, p. 23). One of the difficulties is, however, measuring and understanding the transformation process occurring in the office. Or, more simply put, it is hard to determine or analyze what people do and how tasks may be structured to increase the efficiency and effectiveness of this transformation process. Lieberman, Selig, and Walsh (1982, p. 5) came to the conclusion that, in today's office, there is no structure and no clear and easy way to measure the relationship between input and output. Therefore, they suggest that the office has evolved with little integration, putting it into conflict with the factory, which represents an integrated environment.

Not only are structure and processes important when trying to understand the transformation process in offices. Another major, but often ignored, dimension to help define the office is the social interrelationships that occur there. Offices are really sociotechnical systems that are affected by the environment (e.g., contingency, culture, work space design) and internal processes such as employee characteristics, individual preferences, and social interactions.

For our discussion, an *office* is defined as an open system that processes data needed for providing services and information, and it performs business transactions of various kinds. Several activities are executed by individuals using diverse technologies (e.g., calculators, computers, fax machines, and telephones) to accomplish different job-related tasks *usually* away from home in an assigned physical space within a building structure (e.g., office building). In addition, processes and activities interrelate with, and are affected by, social relationships and environmental determinism (e.g., ILMD, competitors, government regulations, and technology). Finally, office work varies greatly in its centrality to the mission of the organization. For instance, the foreign exchange department is strategically critical to the success of a bank while the communication department's activities have much less importance.

Office Work: Developing a Framework

The above illustrates that two things must be accomplished before achievement of effective EUTM can become a possibility: First, successful study of the technology-induced organizational change process requires a framework to help in assessing the interrelationships between various factors affecting the change outcome. Second, the term *office technology* should no longer be used because what is naturally considered to be office technology is not necessarily confined to the office (e.g., the use of computer-based technology and information systems in manufacturing).

The previous sections use an open system approach to technology acquisition and technology-induced organizational change. Two aspects have been emphasized before: (a) Movement or change in one part of the firm leads in predictable fashion to movement and changes in other parts; and (b) the organization is an open system in that it receives environmental input (e.g., materials, manpower, and capital) and produces output (e.g., products and profit), thus resulting in a system in continuous flux. Hence, technology acquisition and organizational adaptation cannot succeed if these relationships within and outside of the system are not carefully considered and made part of any implementation procedure.

The work discussed so far in this book, and the perspective from which it is viewed, is part of contingency theory, resource-dependence theory, and sociotechnical systems theory. No single theory encompasses all the issues addressed. Nonetheless, the book propagates an open and natural system approach. If one were to identify the rational and natural approaches to systems, contingency theory could be called a rational theory that assumes differences among organizations can be accounted for by studying their formal structures (Hickson & McMillan, 1981; Hrebiniak & Joyce, 1984; Pennings, 1987). Recently, such approaches have been tested with the advancement of the "culture-free" hypothesis, which claims that the relationship between certain contextual variables and dimensions of organizational structure are similar across very different societies (Miller, 1987). However, cross-national research would dispute this claim and suggest that culture can affect organizational structure as well as contextual variables in myriad ways (Bhaghat & McQuaid, 1982; Hofstede, 1984; Miller & Droege, 1986; Swidler, 1986). A question posed by Scott provides a caution against depending solely on a rational theory however: "How can the organization function rationally, given that it is open to the uncertainties of its environment?" (Scott, 1981, p. 131).

In contrast, the resource-dependence and sociotechnical system models take environmental uncertainties into account. These models are natural in that they assume that the firm's exchange of resources with the environment results in the system being in continuous flux as change occurs to allow adaptation to new environmental and internal factors. For instance, the resource-dependence model, as proposed by Pfeffer and Salancik (1978), places great emphasis on the importance of the environment in determining organizational behaviour and chance for survival. An illustration of its validity is provided by those organizations (e.g., most government agencies) that operate in such a highly institutionalized environment that it is more important to conform to externally imposed rules than to produce outputs efficiently (Meyer, Stevenson, & Webster, 1985).

The sociotechnical approach to technology-induced organizational change assumes that the organizational structure is adjusted by taking into account two major aspects of organizational structure, the social and the technical. Research by the Tavistock Group raised two questions of goodness of fit that still apply to today's EUTM — the fit between social and technical aspects of an organization and the fit between the resulting sociotechnical structure and the human characteristics of the people who are part of it both need to be examined (Emery & Trist, 1960). As discussed earlier and illustrated with numerous examples (e.g., Volvo and Fiat), alternative social-psychological arrangements are usually possible within the requirements for the technology to get job-related tasks done. These arrangements are further influenced by the firm's experienced ILMD and environmental constraints such as occupational health and safety laws.

The following sections discuss in more detail various dimensions that are part of a model, trying to further explain office work and technology-induced organizational change. How does this model differ from the ones presented earlier? Figure 13.1 builds upon earlier parts of this book by including not only the contingency but also the cultural, individual, technology acquisition, and structural dimensions in the model. The individual is at the centre of the

Development of a Theory 263

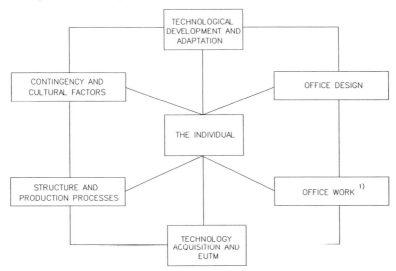

Figure 13.1. Office Work and Technology-Induced Organizational Adaptation: A Conceptual Model
1. Office design and efficiency are two factors that could be renamed "factory design" and "factory work" to apply this framework to manufacturing.

organizational adaptation process. Furthermore, both office design and office work have been added to comply with the need for increased attention to the office domain in any new strategy for technology research and management. Statistics indicate that, in industrialized countries, a majority of today's work force is employed in an office setting. Nevertheless, our knowledge in this area of scientific inquiry is limited, and research and management paradigms must be developed and sharpened to increase our effectiveness. Most important, however, is that the framework in Figure 13.1 can also be applied to the manufacturing sector by relabelling the dimensions: "Office design" becomes "factory" or "shop-floor design" and "office work" changes to "factory work."

Training. Training is a major factor to be considered when trying to achieve effective organizational adaptation to new technology in the office. Acquisition of new technology usually necessitates skill upgrading or changes, as outlined in Chapter 12. As previously defined, *training* in an organizational context is any organizationally initiated procedure intended to foster learning among organizational members. *Learning,* similarly, is a process by which an individual's pattern of behaviour is altered in a direction that contributes to organizational effectiveness (Hinrichs, 1976). Thus the ultimate goal is to achieve greater individual effectiveness by providing the training needed to accomplish desirable behaviour changes.

The changes caused by technology acquisition require continuous training and upgrading of skills to achieve effective EUTM (Okubayashi, 1986). Again, the dimensions outlined in Figure 13.1 affect training. New technology may require new skills, and, if individuals resist technological change, training may help to alleviate some of the employees' concerns. For instance, Gattiker and Hlavka (1989) reported that attitudes towards computers can affect learning performance, but training may in turn change attitudes positively. In other words, training can be used to increase acceptance of new technology.

Even though informal training has been associated with lower levels of literacy and satisfaction (e.g., Bikson, Gutek, & Mankin, 1987), trends show that, during the late 1980s, formal training in the office technology domain dropped while informal training increased (Cooper McGovern, 1988). More research is needed to assess the interrelationships between training and the dimensions in Figure 13.1.

Office design. Research in environment and behaviour, and also in ergonomics, suggests six general office characteristics that might influence quality of work life. Openness is the first one, referring to the total square footage of the office compared with the total length of its interior walls and partitions; the second is density, that is, the number of square feet available to each employee. The third factor is architectural accessibility, the extent to which an employee's individual work space (e.g., desk) is accessible to the external intrusion of others. The fourth is the overall brightness of the office setting, taking into account illumination levels and wall colours, and the fifth is the ambient environment (e.g., heating, ventilation, and air-conditioning). The sixth factor is workplace ergonomics (e.g., furnishing and workplace layout).

These six factors may have varying effects on employees. Dense work settings lead to more frequent interpersonal interaction (e.g., Oldham & Rotchford, 1983). Distractions, however, are lessened by reduced accessibility. Unless noise levels can be substantially reduced and privacy assured, open office environments may not be beneficial (Hedge, 1982). Moreover, the open office design can lead to problems with the ambient environment because men tend to prefer a 20-21 degrees Celsius office environment but women prefer a warmer 23-24 degrees Celsius (Harris & Associates, 1980). This finding has been collaborated by numerous studies in various organizational settings (e.g., Hedge, 1982). The perceived quality of the ambient environment is thus reduced in an open office and may partly explain why, in some countries, employees try to go back to the group office (Ellis, 1984).

Culture may also play a role in determining office design. Locations for offices also vary based on production facilities, design, and structures to facilitate proximity and interaction with production or the most important markets. By the mid-1980s, most North American and Japanese car manufacturers had established design studio/offices in Southern California in an attempt to catch car trends early.

Having offices in different states or countries increases the cultural diversity experienced by the organization, and workplace design must consider cultural preferences by the local employees. Thus, while in an office conglomeration like Manhattan, the preferred office furniture may be high tech, chrome, and futuristic, in the Midwest, dark colours and oak wood may be favoured. In some countries, open-plan offices may represent the tradition (e.g., Japan), and, in others, private offices are preferred (Europe and North America). Insisting on similar office design and furniture on a worldwide basis is, therefore, not advantageous. Instead, office design should incorporate the wishes and preferences of the employees who will ultimately work in the office environment. Cultural diversity is thus encouraged to facilitate creativity and innovation (e.g., Goldstone, 1987).

Office work. The term *office work* includes four relevant aspects of the work itself: the work resources, the environment, the tasks and requirements, and the object of the work. Contingency and cultural factors both influence office work. For instance, educational policy may determine what kind of skills new job entrants bring with them after having acquired a certain level of formal education. This in turn will influence what type of additional training is needed to match the individual's skill levels and knowledge to the actual job requirements. The structure of work and production processes will determine what types of tasks must be performed and the resources available to the job holder; technology acquisition will mean adjustments in work.

In many instances, the machine amplifies the employee's effort, as when the individual uses a spreadsheet and the computer can perform the necessary calculations, thereby helping the individual become more productive. At a more advanced stage, the computer may follow a set of instructions and perform the task without the individual doing the actual work. An example might be a spelling or writing style checker for which the individual is only required to tell the computer what kind of correction to make when an error is found. Both of these represent various levels of mechanization.

A machine performing with stored instructions without human intervention represents an automated process, as when a computer prepares address labels or sends electronic mail to certain parties based on a mailing list. The more we understand about office work, the higher the level of mechanization for simpler tasks will become, and the more likely it is that some of these tasks will be automated.

Most important, however, is the fact that job-related tasks in offices may change in content primarily due not to new technology but to environmental determinism. For example, deregulation of telecommunication in the United States led to reduced long-distance rates, and local phone companies no longer obtained a percentage of these charges. Thus, to sustain profitability and service, technology acquisition was needed to reduce overhead costs. The first area affected was labour-intensive customer services such as directory information and the operator. Now, when calling the information service for a number, the human operator takes the particulars, then the computer gives the customer the phone number requested. Computerization has also helped to reduce the human work involved in responding to the common complaint of losing change when using a public phone. Today, when someone calls for a refund, as soon as the operator has decided that the person deserves money back, the computer takes over, asking the customer for such information as how much money was lost, phone number of the public phone, time, date, and address where the check should be mailed. The computer then informs maintenance of a potential malfunction that should be repaired and makes sure that a check is processed and mailed. These two examples illustrate how office work can change by partly mechanizing (information) and automating (refund) labour-intensive customer services. Some phone companies chose to automate both information and refund procedures. However, they had to maintain personal service as an option for customers who had gotten confused by the new technologically oriented system and refused to be forced into using it.

Figure 13.1 shows the various interrelationships between an office and other factors such as individual characteristics, office design, and technology acquisition. Accordingly, changes in any of the dimensions in Figure 13.1 will ultimately lead to adjustments in one's work either directly or indirectly through another dimension. For example, the type of office work and an individual's characteristics can affect the type of technology acquisition schedule (radical versus incremental). But in turn technology acquisition may lead to changes in the office job that indirectly affect the employee (e.g., upskilling versus deskilling; Attewell, 1987a; Spenner, 1983). Environmental determinism may also force the firm to change its EUTM strategy to adapt

Development of a Theory 267

to new conditions (e.g., deregulation). This indicates that the office job itself is just one dimension that affects and is affected by other factors in the framework outlined in Figure 13.1.

Technology acquisition and EUTM. EUTM with equilibrium between financial, technical, societal, and human constraints, and with a focus on the individual end-user, is necessary for constructive use of technology. At the acquisition and adaptation stage, contingency and culture, and also the structure and processes of production, are extremely important. For instance, ILMD may lead to resistance regarding certain technological developments. Often employees perceive technology as automating the office, which does, of course, lead to job losses and other undesirable outcomes (Doswell, 1983, chaps. 1-5). However, only well-understood tasks can be automated and, because most tasks involved in office work cannot, mechanization, instead of automation, is more likely to occur (Doswell, 1983, chap. 4). If employees believe this to be true, resistance may be decreased.

Gender differences can also exist (see Chapter 11), although recent work has argued that their cause is not always clear (Jacklin, 1989). For example, although research indicates that women are more easily trained (require less time) than men when acquiring skills, an explanation has not yet been forthcoming to account for this phenomenon (e.g., Gattiker, 1990a). Hence, gender differences may primarily be based on sociological factors (e.g., women taking less technical training during schooling — Lowe & Krahn, 1988) and females' lower expectancy for high performance with technology (see Vollmer, 1986).

When introducing new technology, successful management of the organizational adaptation process is naturally also influenced by contingency and culture. For instance, market conditions may affect the firm's choice in this area. Moreover, organizational culture may require a certain approach to assure acceptance of the change by employees. The organizational structure will in turn influence how technology can be integrated. In most instances, technology must fit the current structure and not vice versa (see Emery & Trist, 1960).

If high effectiveness appears to be achieved through the introduction of new technology, the firm may be encouraged to push ahead and introduce even more advanced technology in the office. This may at first seem rational because success invites management to continue along the same path. Nonetheless, there may come a point at which the employees neither welcome nor accept such change but instead resist it fiercely. The manifold reasons for this were outlined in Chapter 11, so they need not be repeated. Suffice it to say that the interrelationship is extremely complex and must be carefully studied and assessed on a continuous basis to assure a smooth transitional period from an older technology to a more advanced one.

Effectiveness has become even more important because costs have soared. Recent figures show that the organizational investment for an office job sometimes surpasses that needed for a factory position. Office space in some cities is so expensive that various firms have decided to move part of their offices outside of downtown locations (e.g., Citibank relocated its credit card operation out of Manhattan, Academic Press moved some operations from New York to San Diego and Tampa). The full liberalization of trade between member countries of the European Community planned for 1992 (financial markets will follow a bit later) will offer similar opportunities to firms in Europe. Personal computers and other office technologies will accelerate this trend as communication and data transfer enable the firm to locate in geographical areas where real estate is still relatively inexpensive, thus reducing office overhead despite substantial initial investments (Nolan, Norton & Co., 1987; Treacy & Index Group, Inc., 1989).

I will reiterate that effectiveness in the technology acquisition domain is in flux. What might be effective today may be ineffective tomorrow. Changes in labour laws or deregulation of a market may force the firm to totally rethink its technology acquisition and HRM strategy to meet the competition head on and with success. Unfortunately, static models have been used to study EUTM that do not consider all the necessary factors. Instead of cross-sectional designs, time-series studies comparing the dimensions over several points in time are needed to improve our understanding of the process required to accomplish effectiveness in technology acquisition. Figure 13.1 outlines a model that is of potential benefit to employees and organizations and thus warrants further testing in a time-series approach.

Structure and production processes. As discussed in Chapter 3, the structure of an organization affects its production process. For instance, mass production requires different types of jobs in the office than multiple product lines, which may require different accounting procedures, customer service, and production facilities for each product line. Chapter 9 outlined how production is organized differently in various countries (e.g., France having more supervisory personnel than West Germany—see Maurice, Sorge, & Warner, 1980). Other differences are based on legally stipulated participation by workers when introducing changes in production processes. This leads to formal influence by workers in some countries (e.g., Wilpert, 1986), but in many other nations, worker participation in the decision-making process when considering potential development and acquisition of new technology is either on a more informal basis or nonexistent for all practical purposes.

In Japan, nonunionized firms rarely obtain input and suggestions from workers about imminent work and structure changes due to technology (Okubayashi, 1987).

Chapter 3 also outlined the differences between a U-form and an M-form organizational structure. The M-form facilitates participation and an especially quick response to technological developments. The higher financial returns for organizations using the M-form (Burton & Obel, 1980) may make it easier for management to achieve technological effectiveness. Nonetheless, this line of argument is weak in that it assumes rationality on the part of individual participants. Thus, even though state-of-the-art technology has been installed, human beings may still resist using it. Facilitating worker participation does not rule out resistance. Resistance may be grounded in various fears, such as anxiety that lower management positions will shrink as a percentage of overall employment. There could be status-conscious concern that the manager's ultimate symbol of prestige and power, a private secretary, will become extinct as the effective use of computers permits sharing of support staff (Zuboff, 1982).

Status-conscious managers are also unlikely to personally use expensive workstations to any great extent, because doing so smacks of clerical work. This effectively eliminates the potential time savings on which basis the investment was made in the first place. Thus the workload for the secretary is not reduced and a personal secretary remains a necessity. However, the most severe drawback is that utility is reduced, because, in addition to the technology investment, labour savings and productivity increases will not materialize. Technological effectiveness may also be lowered if the technology does not bring qualitative and quantitative improvements simultaneously. To accomplish increased technology use by higher echelons, the strategy most likely to succeed would be for the chief executive officer to share his or her private secretary with another executive and improve his or her level of technology skills to provide the basis for more extensive computer-mediated work as an example for the rest of the firm. This may be difficult to accomplish as long as a stigma is attached to typing that encourages executives to shy away from computer keyboards (Morgall, 1983). In 1989, West German firms providing executive training seminars for computer skills believed that executives only needed to understand the concepts behind information systems (Lerne und arbeite, 1989). The gospel was that procedural knowledge about using WordPerfect or Lotus on a computer was not required for an executive because one's secretary would know about these programs and perform tasks necessitating their operation.

How does the above relate to organizational structure? Even though it may appear that the M-form facilitates technology acquisition, people's behaviour and attitudes within the organization may still limit the outcomes. As a consequence, structural advantages may fail to materialize due to irrational employees or managers who do not want to give up their status symbols. Again, the individual is the key to finding any solution. Even though radical change in an M-form organization may be more advantageous, ILMD, worker resistance and managerial behaviour may result in the less advantageous incremental organizational adaptation instead.

As illustrated in Table 4.1, ownership, governance, and distinctive competence also affect how the firm will approach technology acquisition and organizational adaptation. Additionally, organizational culture may suggest that an incremental change is better than a radical one. For instance, with the deregulation of air traffic in the United States in the early 1980s, some carriers reduced their ticket prices and tried to radically adjust to market changes (e.g., Texas Air Corporation), but others used their technology and seat management systems to adjust pricing and scheduling to increase their revenue per passenger mile and on-time performance. In this case, the incremental change proved superior. American Airlines and its aforementioned SABRE system are an example of such an incremental change accomplished through the constant upgrading of the computerized reservation system. About 60% of all ticket sales are done by travel agents using such systems and, because each system lists the service providers' flights first, the majority of bookings are made using the service provider as a carrier.

In an attempt to imitate this success, Texas Air (which already owned Continental, New York Air, Frontier, and People Express) took over Eastern Airlines to increase its market share in the United States and, most important, to acquire Eastern's reservation system. Texas Air was hoping to gain the same advantages enjoyed by American Airlines by offering such a system to travel agents and benefiting from it in various ways (e.g., increase agent's percentage of flights booked with Texas Air or collecting fees for reservations made on other carriers via the system). Moreover, the technology was supposed to facilitate seat management and thereby increase revenue per passenger mile. A quantum approach was used to reorganize Eastern, but labour unrest and other difficulties such as flight cancellations and delays resulted in a $466 million loss in 1987 for Texas Air while its competitors were making hefty profits. One reason for the labour unrest was a clash between cultures, because Texas Air's autocratic and aggressive management style (e.g., demanding pay cuts from employees and increased workloads as not all departing employees were replaced) was not appreciated by Eastern Airline's workers, who were used to a different type of management. In this case, the takeover and the quantum approach used to reorganize resulted in labour problems that offset the benefits obtained through the acquired reservation system and ultimately led to the bankruptcy and breakup of Eastern Airlines in 1989. Thus internal growth and technological development of a reservation system from scratch might have left Texas Air with a lower debt, fewer labour problems, and more profitability. This is just one more example suggesting that, although technology acquisition may depend upon numerous factors, the individual's acceptance (worker and customer) of change is the crucial ingredient for success.

Contingency and cultural factors. As earlier parts of this book have outlined, contingency and culture affect the firm's technology acquisition and development strategy. For instance, ILMD affects the changes that must accompany technology introduction; often before technology can be adopted into the workplace, job descriptions must be rewritten and necessary architectural changes in the office should be made.

As discussed in Chapter 10, certain actions and behaviours may have specific symbolic meanings in an organization. Cultural values may be made explicit through office design by various architectural means including interior layout or furnishings. These indicators may be blatant or subtle. The "corner office" going to the person highest in the hierarchy is just one example. Companies may or may not allow the personalization of work space (e.g., family pictures on desks), and colour schemes can be indicative of professional groups (such as panelled walls in law offices and bright colours in pediatricians' consulting rooms).

In West Germany, as mentioned earlier, more and more unions and organizations have agreed to change back from an open office design to the concept of *Gruppenraum* (group room) to increase the group's privacy, control over heat, lighting, and other services, thereby improving the work environment (Ellis, 1984). Again, these changes developed out of a culture and ILMD that allowed workers to ensure that their wishes were heard, which affected the decision-making process about office design, using both formal and informal means of influence. In contrast, in North America ergonomic concerns about office design are seldom raised using formal channels, and employees are usually faced with facts but not given the opportunity for input during the early design stage. It could also be argued that turnover and mobility are factors that may have more impact on the considerations of North American companies. Research shows that many employees under 40 change jobs with some frequency (see Carter, 1988; Hall, 1982). Management may, therefore, make work space decisions without consulting workers who might have left by the time the changes are implemented.

Office design is also influenced by environmentally determined factors such as occupational health regulations, but effects will differ based on national culture. For instance, in Sweden, the potential health hazzards connected with using visual display units (VDTs or computer terminals) have become a major concern for governments and unions, resulting in numerous laws and ILM regulations aimed at preventing undesirable effects. The Central Organization of Salaried Employees (TCO) in Sweden has published the Screen Checker, which allows the employee to check computer terminals for ergonomically "friendly" qualities such as readability, height and possible tilting of screen, and many other features facilitating work with a terminal. The employee is also advised to talk to his or her union representative or the organization's health care services about the latest research results for potential radiation effects upon skin and pregnancy and other health concerns (TCO, 1986). In contrast, North American workers lack this protection or advice. Also, their awareness of health and ergonomic issues and new technology may be lower than that of their Swedish colleagues, suggesting some cross-national differences. Although in Sweden pregnant women's exposure to radiation from computer terminals must be reduced by law, such a desirable scenario is still far in the future for the North American employee. Additionally, workers' councils and ILMD give Swedish workers the means to influence the design of their workplaces while in other countries participation is either voluntary or more often nonexistent (see also the section on office design in this chapter).

The individual. Attitudes towards technology, needs, social comparison processes, and perception of equity and equality are included here. For instance, based on recruiting and selection strategies, the organization may primarily have hired individuals who already possess some technology-related skills when joining the firm. It is likely that these individuals feel comfortable working extensively with office technology on a daily basis. A recruiting and selection strategy that tries to encourage people with desirable skills to apply for employment and to take a job when it is offered represents a conscious choice made by the firm. Such a choice in SHRM affects not only today's potential technological effectiveness but also the firm's future capability for technology acquisition. Labour market power, which allows the firm to attract individuals with the required skills, is accomplished by setting wage rates above the market rate (Windolf, 1986), and while this power facilitates recruiting, a thorough and systematic selection system allows the firm to hire people with the necessary declarative and procedural knowledge to promote technology-induced organizational change.

Labour market power also facilitates a firm's explicit or implicit attempt to hire people with the proper technology disposition (see Gattiker & Nelligan, 1988). Thus the individual possesses relatively stable traits (e.g., beliefs and needs; Rokeach, 1980) that positively influence his or her affective (i.e., attitudes) and behavioral reactions towards technology-induced organizational change. This assumes that individuals have a predisposition towards technology, which significantly affects their acceptance of technology-mediated work in organizations (Staw & Ross, 1985). Although dispositions can change over time due to organizational and work-related influence (e.g., Brousseau, 1978; Kohn et al., 1983), a firm may prefer to hire the people with the "right" attitudes instead. This approach is viable for some firms, but other institutions cannot choose their participants (e.g., schools, voluntary organizations, rehabilitation work centres and prisons). In such situations, people require support through training to help them adjust to new technology (Gattiker & Hlavka, 1989) and change their dispositions towards technology-induced organizational change (Davis-Blake & Pfeffer, 1989; see also Table 12.1).

Highly competitive remuneration packages will attract qualified employees to work for an organization; however, long-term employees doing the same job may be paid less than newcomers. A strong ILM usually leads to external inequity over time (i.e., long-term employees are paid lower wages than is the industry standard). From a bystander's point of view, this also represents inequality within the firm. Two possible scenarios may emerge. The firm could simply increase the salary of the more tenured employees to reduce or even eliminate the inequality, but in many cases this does not happen. Instead, the firm may hope that the quality of work life and other nonmonetary factors may compensate for the internal inequality to assure the further participation of senior workers.

Research Approach

Figure 13.1 outlines a framework that should help in further developing the paradigm for office work and technology-induced organizational change. To begin with, some principles of theorizing must be discussed. Thereafter, theoretical statements or hypotheses to be addressed in future studies are outlined. These theoretical statements are based on the literature already reviewed here. This literature represents a vast body of research compiled over the last few decades, suggesting that we have come a long way to explain many puzzling phenomena pertaining to technology acquisition and organizational adaptation. Nonetheless, the list of possible hypotheses to test is rather long yet by no means complete. It merely represents an appropriate reflection on the topics discussed in this book. Most readers will be able to add to the lists, depending on their areas of expertise and interest.

Principles of Theorizing

The requirements of deductive theory can be subsumed under four main principles (see Blau & Schwartz, 1984, chap. 1; Hall, 1982, chap. 7). First, a theory is composed of propositions that refer to the relationship between two (or more) independently defined concepts (e.g., *skill acquisition* is the process of gaining procedural and declarative knowledge — see Chapter 12) and are thus subject to empirical testing. Second, the propositions constitute a hierarchic, logical system in which all lower-level propositions can be deduced from higher ones (*theoretical definitions* and *linkages*, using Hage's, 1972, terminology). If a proposition is not deducible, it represents an assumption, or postulate.

A third characteristic of a deductive theory is that it contains some purely theoretical terms that cannot be empirically measured or that only summarize empirical findings logically implied by them but do not provide new ideas to explain these findings (Braithwaite, 1953). Homans (1980) argued that some propositions cannot be deduced from others but are invented to facilitate our understanding of a phenomenon (e.g., Newton's theory of gravity). The fourth characteristic is that theoretical propositions must be sufficiently precise to be translated into operational variables for research, which provides the opportunity to test the theory (Hage, 1972, chap. 7).

Before we can develop the theoretical propositions and operationalize the variables, however, we need to determine what the dimensions to be used in this theory will be. It seems that, based on the discussions in this book (see also Chapter 2 and Figure 2.1 for an overview), two dimensions must be given particular attention by employees and firms alike: The focus of study must include *individual* (micro) and *organizational* (macro) levels of analysis and should span the *objective-subjective* dimension of inquiry when studying the technology management process.

Another distinguishing factor for the approach outlined in this book is that most recent models for technology management have concentrated on explaining the necessary processes for creating an environment for technology (e.g., changes in production facilities and office design) with a limited focus upon the human component and its influence on potential success in this area. Most office design research has not been cross-fed into the organizational behaviour domain or ergonomics. Thus architects, engineers, psychologists, sociologists, and others are each going their separate ways in analyzing technology's effect upon office work. A fully comprehensive strategy is lacking, and reasons abound. Primarily, it is virtually impossible to find one scientist who is knowledgeable in all of these areas of research. Multidisciplinary research is, moreover, difficult and time-consuming, so academics avoid it because they are evaluated in part on productivity over the short term (e.g., number of published articles). And the realization of such a research project might take up to five years from design to write-up, involving experts from various disciplines and requiring much patience from all parties involved. Further, it may not result in any publishable work for up to three years. Unfortunately, today's academic reward system does not allow for such long dry spells. Quantity is often more helpful in advancing a researcher's career than quality. Although these difficulties are understandable, they do not justify the lack of such efforts.

Instead, our knowledge about the effects of technology acquisition will remain limited, and social impacts will neither be assessed nor predicted with any certainty, if such multidisciplinary, long-term projects are not encouraged and financed by both business and government. The overall research paradigm that is so vitally needed will not be developed until these kinds of studies are conducted. With everyone using such different research approaches and methodologies, and such widely varying definitions of constructs, it is no wonder that rapid progress has not occurred. Unless we understand the way that office jobs are affected by the dimensions outlined in Figure 13.1, our understanding of technology acquisition and its effect upon office work will be limited.

Theoretical Statements

Office work and technology-induced organizational change have occasioned large bodies of both empirical and theoretical literature, which are rarely joined. Furthermore, theoretical formulations are hardly ever used to guide empirical research (see Blalock, 1984, chap. 7). Let me attempt to develop some research questions based on the frameworks presented earlier in this book (e.g., Chapter 2) and on Figure 13.1.

Before, however, on a more conceptual note, there are numerous names and labels for theoretical statements; one may make an *assumption, hypoth-*

esis or *premise* or state a *theorem*. *Hypotheses* are assumed to be unconfirmed theoretical statements, and *propositions* are assumed to be substantiated by evidence. Because there are so many names, Hage (1972, p. 35) suggested the use of the more neutral term *theoretical statement*, although here the term *hypothesis* is sometimes used as a synonym because it has the most widespread usage in organizational science, psychology, and sociology.

The theoretical statements listed below are based on the material discussed in this book. After each statement, the reader is given the appropriate chapter numbers to allow further study.

Objective Dimension of the Framework

Based on Figure 2.1, the objective dimension of the framework deals with structural issues, ILMD, choice in SHRM, and technology and general skill requirements at the individual level. Moreover, the statements integrate the issues addressed in this chapter (see Figure 13.1).

Macro focus. Some of the hypotheses to be investigated are as follows:

(1) Technology acquisition and technology-induced organizational change are facilitated by high choice in SHRM (Chapter 3).
(2) Technology acquisition and technology-induced organizational change are facilitated by limited ILMD (Chapter 5).
(3) Rigid hierarchic organizational structure reduces the adaptability of the firm (Chapter 4).
(4) Multiple production processes facilitate an organization's technology-induced adaptation process (Chapter 4).
(5) Innovative and systematic recruiting efforts increase a firm's labour market power (Chapter 6).
(6) Systematic selection procedures and career development efforts increase the labour market power of the firm and increase successful employee learning in company-sponsored training programs (classroom or on the job; Chapters 6 and 8 as well as 12).

Micro focus. Figure 2.1 indicates that technology and skills fall under a micro focus as part of the objective dimension. The questions to be researched in this area are these:

(1) Systematic socialization and appraisal procedures increase job performance (Chapter 7).
(2) Training given by the firm provides more procedural knowledge than that given by educational institutions (e.g., college, computer dealer); conversely, out-of-

Development of a Theory 277

house training provides the individual with more declarative knowledge. Thus external training is more transferable to another job and/or employer than internal training (Chapter 12).

(3) Intermittent training (e.g., one day for several weeks) results in higher learning performance than continuous training (Chapter 12).
(4) Accessibility of technology increases its use (Chapter 12).
(5) Increased use of technology augments the technology skills held by the employee (Chapter 12).

Subjective Dimension of the Framework

The subjective dimension of the framework is represented in Figure 2.1 as culture, individual perceptions of technology-induced organizational change, social comparison, and acceptance of new technology in the workplace.

Macro focus. Some of the theoretical statements needing investigation are as follows:

(1) Cross-national differences in culture and work values result in different office design and organizational structure and varying technology-induced organizational change processes (Chapter 10).
(2) A strong organizational culture provides a structure for appropriate action and behaviour when adopting new technology and making the necessary organizational changes (Chapter 10).
(3) The more rigid the organizational hierarchy, the more standardized office design will become, with variation and individuality being limited to a few personal items (Chapters 4 and 11).
(4) If the organizational hierarchy is inflexible and ILMD is high, office space and furnishing will depend greatly upon one's hierarchic level (Chapters 4, 5, and 10).
(5) The greater the ILMD, the greater the participation by employees or their representatives occurring during the decision-making process about new technology (Chapter 10).

Micro focus. Some of the hypotheses that evolved out of the present discussion are as follows:

(1) Individuals' perceptions of inequality increase resistance towards new technology (Chapter 11).
(2) Increased ILMD reinforces the employees' emphasis on equality (Chapters 4 and 11).

(3) Low ILMD adds to the variety of office work performed by employees on their job (Chapters 4 and 13).
(4) Low ILMD and high choice in SHRM increase inequality in remuneration and reward-based compensation packages (Chapters 4, 5, and 9).
(5) Systematic individual career development eases technology-induced change as experienced by the employee (Chapters 8 and 11).
(6) Systematic training efforts increase acceptance of technology-induced change by the employee (Chapters 11 and 12).

Implications and Future Strategies

As previously stated, research efforts need to become more longitudinal, and the methodologies used must include surveys, observations, interviews, and other methods to attain a comprehensive picture of the relationships between the dimensions outlined in Figure 13.1. Moreover, a more dialectic approach to research is required; that is, the merits of opposing views must be tested. Failing to reject hypotheses may be as useful as rejecting them if doing so increases our understanding of technology acquisition and management processes.

The theoretical statements outlined here are precise enough to allow for testing. They are based on the dimensions (macro-micro, objective-subjective) of a hierarchic, logical system in which the lower-level (micro) propositions can be deduced from the higher ones (macro). These theoretical statements are by no means exhaustive; they merely lead to possible avenues of study, providing a starting point. Future work must build upon the rudimentary framework presented in Figures 2.1 and 13.1.

Summary

I must emphasize again that a higher level of paradigm development is needed to advance our knowledge about technology acquisition and technology-induced organizational change. A well-developed paradigm will help the discipline find a common ground of understanding, thereby providing some guidelines for appropriate research endeavours. The theoretical statements outlined above only represent a small number of possibilities. Nonetheless, they succinctly outline what issues must be addressed based on the empirical data and findings discussed in this book. The testing of these hypotheses within the framework developed here should propel paradigm development in the technology acquisition and organizational adaptation domain beyond its current rudimentary stage. Furthermore, resolution of the issues presented above will ultimately increase our knowledge about EUTM and possibly increase management's effectiveness in this area.

More specifically, this chapter describes some of the vast changes taking place in offices, which are the workplaces of an ever-increasing number of employees. The office as a unit of analysis must be investigated more thoroughly to help us better understand technology-induced organizational change. Hence, a well-developed paradigm for the office is needed to facilitate our future research endeavours in this area. Knowledge in human relations, financial and accounting issues as well as ergonomics, organizational design, and architecture, to mention a few disciplines, is advantageous when doing research about technology in an office setting.

14. Conclusion

This book's objective has been to examine a number of technology acquisition and organizational adaptation issues that are shaping the development and functioning of organizations. Focusing on human resource management concerns when discussing innovation and technology-induced organizational change is a relatively new method of inquiry. There are still substantial gaps in empirical data regarding both human resource management and choice as well as ILMD on the extent to which they interrelate and how they are managed. Indeed, this book raises as many questions as it answers.

The primary motivation for this work arose from a desire to provide a more balanced treatment of technology acquisition and organizational adaptation from an end-user perspective than is currently available. Most of the existing studies treat the subject technically, concentrating, for example, on the innovation process itself (e.g., Kamm, 1987) rather than on its results. Another limitation of previous work has been its focus on organizational concerns while practically ignoring the fact that human resources are an important asset in the management of technology acquisition and effective EUTM. Exceptions do exist, but even these concentrate on human resource issues in high-technology firms, ignoring the fact that the vast majority of firms that must deal with technology acquisition and especially organizational adaptation are not in the high-technology domain per se (e.g., real estate, finance, health care and construction firms). Such treatment is often very superficial, naive, or anecdotal and generally lacks any conceptual or theoretical framework (with some notable exceptions, such as Hirschheim, 1985). However, theoretical and methodological concepts must not be disregarded when trying to understand adaptation processes and effective EUTM. Rather, philosophical considerations should provide the foundation for all inquiry.

This point is extremely important, and a good part of this book has been devoted to defining and outlining the conceptual and theoretical notions underlying the contingency and cultural approaches to technology acquisition and organizational adaptation. The social theoretical approach, adopted

Conclusion

here by combining the two, has a firm foundation in the more "interpretivist" philosophical camp, based on the belief that, as the unit of analysis is the individual, positivistic concepts of inquiry are inappropriate. People have great will and possess an intersubjective conception of reality. Their values, beliefs, and systems of meanings, as well as their interpretation of situations such as the work environment, may sometimes appear irrational to outside observers. Thus assessing technology acquisition and organizational adaptation from a systems point of view utilizing the rational approach is simplistic and leaves too many questions unanswered.

How does using the combined contingency and cultural approach to study these change processes help avoid the problems incurred with other approaches? Technology acquisition is a process occurring in social systems, mainly organizations. Also, technological innovation has been defined not as being part of a technical system that has social implications but as a social process of a social system that increasingly relies on strategic HRM for its effective operation. It is this social system orientation, and especially the human resource focus, that is lacking in most treatments of technology acquisition and related organizational change. It is difficult to understand the processes within the social system and its exchanges with the environment without considering both the contingency and the cultural approaches.

It is not surprising that the technical side of the "man and machine" equation has received considerable attention. The process of innovation and the technical aspects of information technology are exciting; managers and other stakeholders see in them the potential for increased efficiency and profit. Vendors and consultants see a huge market for their products and services and want to point out the economic benefits to potential customers. Academics have by and large also stressed the innovation process itself and the technical side of technology management (see especially the information systems literature), often with reference to high-technology firms. One reason for this has certainly been governments' tendency to put innovation and technological competitiveness on their reelection banners. Hence, funding for innovation, and research on the subject, has been substantial.

Yet some research suggests that "high-technology" firms do not create more employment than "low-technology" organizations. In fact, employment may be reduced with increased innovation (e.g., Gibbs, Keen, & Lucas, 1988). Therefore, various location subsidies offered by government agencies for "high-technology" firms to establish branches in more rural areas (e.g., where high unemployment exists or where traditional industries, such as mines, may have been closed down) may not result in the desired economic growth. Three reasons can be identified: First, former mine workers may not have the necessary skills to become employed in a "high-technology" firm; second, these firms may not be as labour intensive as mining; and, third, much of the "high-technology" production work pays low wages compared with those offered in traditional industries that are often unionized.

Numerous researchers have tried to point out the social impacts of new technologies on work and organizations (e.g., Bikson, Gutek, & Mankin, 1987). Many studies have tried to assess structural and environmental effects on organizational performance and technology acquisition (e.g., Hrebiniak & Joyce, 1985; Young, Hougland, & Shepard, 1981), and others have evaluated attitudinal and behavioural outcomes (e.g., Floyd, 1988; Gattiker & Nelligan, 1988). Both of these approaches are important, but we must strive for a combination of contingency and culture (values, needs, habits and action strategies). We should know why individuals feel a certain way about technology or why an organization introduces a technology later than its competitors. An integration of both perspectives for a more thorough study of technology acquisition and organizational adaptation processes will help our understanding of the process, its outcomes, and its impact upon human resources.

It is tempting to suggest that the growing review and reflection, coupled with ambiguity, is indicative of a paradigm crisis in technology acquisition theory. If so, such a crisis is definitely needed. As noted at several points throughout this book, the dominant perspective on action that has characterized technology acquisition and most innovation research has been the rational model, which studies the choice and determinism dimensions as focused upon by management. Our review of this approach in Chapters 3, 4, and 13 makes the limitations of this method clear. The underlying theoretical premise of the rational perspective—that individuals do things because they want to—has grown increasingly troublesome because of its inherent circularity (see also Part III).

In later parts of the book, the focus has been mainly on the office and information technologies—one area that has experienced tremendous innovation and technology acquisition and will probably continue to do so. The proliferation of personal computers in office settings is ample evidence of this trend. Yet too much past research on technology acquisition has concentrated on the factory while acquisition and organizational adaptation in the office domain has had too little attention paid to it. Continuous new developments in the personal computer area require firms and researchers to keep abreast of these changes that benefit office employees more than factory workers.

Office technology has changed tremendously just within the last decade, mostly due to the advancement of PC technology, which made it more accessible to a nontechnical end-user. Today, many employees have access to and do part of their daily work using computers, but knowledge about the organizational adaptation process in the office technology field is very limited. One thing, however, is certain: The individual is the most important factor to assure EUTM effectiveness. If a person accepts computers, this attitude will most probably, but not inevitably, increase his or her use of the technology. People may accept technology but still resist using it, as research has shown (e.g., Dierkes & von Thienen, 1984). One likely reason for resistance may be the sometimes negative media attention given to computers and also possibly the anxiety and fear the employee feels towards this technology because it may threaten job security and lead to undesirable work changes. Although rational analysis of the situation should indicate to the employee that computerization will help ensure the firm's competitiveness, thereby increasing long-term job security, assuming rationality in this debate is futile as individuals are seldom truly rational. Based on their limited information, employees may conclude that the best strategy is to resist technology-induced change (e.g., Simon, 1979). In the past, however, a rational approach in technology management has been assumed, leading authors to make assertions and conclusions that are rarely helpful.

What I hope the reader has grasped from the review of the approaches that emanate from such a perspective is how difficult the rational approach in technology acquisition is to operationalize. It may fit some cognitive biases we share about individuals and organizational systems, but it has yet to demonstrate the ability to advance technological analysis beyond the limited knowledge already available. The current state of the field provides evidence of this problem.

Synthesis

Chapters 2 to 4 outlined how environmental determinism, organizational structure, and ILMD may together affect the choice in SHRM, which ultimately influences the company's strategy when it comes to technology acquisition and organzational adaptation. The approach used went further by not only outlining how external constraints (competitors, regulations, and innovation) may affect organizational choice but, equally important, how internal constraints such as ILMD affect choice in SHRM. The latter in turn may improve the choice experienced by the firm when it comes to environmental determinism. In the past, few researchers have pointed out that ILMD may affect choice in SHRM and thus possibly reduce the competitiveness of the firm and its choice of when and how technology acquisition will occur (e.g., Gattiker, 1988a).

In this book, the focus has been on how the HRM component, ILMD, and the structure and processes of the firm might affect choice in SHRM. For instance, the organizational types outlined in Chapter 4 suggest certain organizational systems that affect technology acquisition either directly or at least indirectly. This has been discussed in detail in Chapters 5 to 9. The subsystems described — such as selection, recruitment, compensation, and career development — together influence the type of individual hired by the firm and the values and action strategies employed by its current workforce. Hence, the systems used by each of the four organization types help attract and retain certain individuals who share beliefs and values and give similar meanings to actions, thus forming the core of the organization's culture (Chapter 10).

The cultural approach also requires a look at the social comparison process and workers' subsequent acceptance or resistance towards technology acquisition in the workplace (Chapter 11). This in turn may affect or at least relate to the level of technology skills attained by a firm's work force, as discussed in Chapter 12. For instance, employees who feel unjustly treated in comparison with others when it comes to new technology may resist an organization's attempt to introduce more advanced technology into the workplace.

I hope that the book, as summarized above, illustrates how critical it is for today's organization to examine technology acquisition from a multidisciplinary approach, looking at systems, contingency theory, human resources, and many other factors to ensure effectiveness, which in turn justifies the huge investments required, in both human and financial resources, to implement the acquisition of new technology (Chan & Mountain, 1987). The analysis and approach in the preceding chapters lead to an inescapable conclusion about the fundamental issues in technology acquisition. It is of crucial importance to advance our knowledge in the technology acquisition domain with all due speed. The literature in this field, as a whole, has tended to move too far from the data and findings. Put differently, it seems there is a lot of ideology and assertion but not enough attention to the results, or lack thereof, from the various empirical investigations that have been undertaken. There is a real urgency now to turn back to the basic properties of human resource management and technology acquisition in conjunction with organizational adaptation so that our understanding of these issues will truly improve.

Research Implications

In Chapter 1, the reason for this work was discussed in terms of the need for a treatment of technological innovation from a human resource perspective by giving priority to contingency, cultural, and social rather than technical aspects of the subject. This has been undertaken through the application of the human resource perspective, which has proved valuable in its ability to generate insight into the social nature of technology acquisition. It is likely, however, that the conclusions reached by this study might seem either obvious or irrelevant to some — obvious because "everyone already knows" that the human resource factor is important in managing technology acquisition and irrelevant because we have already managed to adopt technology without such research.

The first criticism is one suggested with some regularity in response to most, if not all, work done on the social and human resource management aspects of technology acquisition. It is raised by "enlightened" individuals who believe that technology acquisition needs to be addressed through research on the innovation process itself or on the man-machine interaction. Because such interaction has been duly considered, the importance of the human resource side of the issue is felt to have been recognized and addressed adequately.

Yet this criticism is fairly easy to refute. Although it is clear that some progress has been made on the social aspects of managing technology acquisition, this progress is minimal and probably misguided. First, man-machine interaction is not the only thing involved. The human resource dimension involves social interaction — people engaging in various actions and behaviours. The meanings associated with social action are intersubjectively shared. Man-machine interaction processes are acknowledged to be psychologically driven, but this does little to address how the social environment, environmental determinism, culture, ILMD, and structure of the firm interrelate with technology acquisition. Moreover, even though user involvement is sometimes assured, it is usually quite limited. Users may only be involved at the final phase of the decision-making process and thus have little say in the early stages when shaping of the technology acquisition and technology-induced organizational change takes place. Actually, a sociotechnical approach (see Chapter 13) is needed to consider structural and human characteristics and to obtain a fit between the two (Emery & Trist, 1960).

Additionally, technology acquisition is often sprung on the employee without notice, opportunity for input, or even suitable training for its use.

Moreover, a fully integrated assessment of the factors outlined in this book is usually neglected so that a less comprehensive approach (e.g., financial assessment, projections that rarely materialize) is often all that is used to assess the acquisition of and organizational adaptation to technological innovation.

The second criticism — the lack of practical relevancy — is perhaps more difficult to refute. Critics may claim that technological innovations have had considerable success penetrating most areas of business life. So, what do we need to study? Nonetheless, a fuller understanding of the organizational adaptation process caused by technology acquisition and its interrelationship with other aspects of management shows that many public and private actions affect the overall success of technology management.

Practical Issues for Major Participants

Technology acquisition poses both challenges and opportunities to all who are involved directly as well as society as a whole. To capitalize on these opportunities and reduce negative consequences, enlightened actions based on thorough knowledge by those concerned is crucial for, and possibly the only way to, success. Here we focus especially on the practical side of the issues addressed in this book.

Educational Institutions

Most training is provided at the beginning of one's career even though technological change and organizational adaptation require continuous training and upgrading for some employees during their work lives. Postsecondary institutions in North America educate increasing numbers of nontraditional students. Such individuals may have finished high school and then begun regular work for several years, or attended university for a year and then left for several years, returning later to finish their education and receive a degree. Moreover, many of the students may study part-time while working either part- or full-time.

As pointed out in Part III of this book, continuing education programs offer a variety of noncredit courses, but the quality is not always comparable to degree courses. Often these programs are offered by U.S. state universities, in part to serve their public mandate, but the faculty teaching in these programs is usually hired ad hoc, with regular faculty rarely involved. Efforts in this area at the two-year-college level do not always fare much better. In Canada, these programs are limited to managers, while, for example, in Switzerland, recurrent education options through nondegree programs are quite numerous, as employee demands for such programs are extremely high in comparison with other countries (BIGA, 1988; see also Chapters 10 and 12).

In the past, education has been for both knowledge and training. Increasingly, however, the trend is towards professional programs (e.g., business, engineering, nursing, and education), where students are mostly taught skills and techniques while critical thinking and comprehension of complex social and scientific phenomena (procedural knowledge) may take a backseat. Hence, universities are confronted with many students attending merely to obtain a piece of paper to facilitate their access to a prosperous livelihood. The biggest challenge for educational institutions is thus to meet the tremendous pressure for better performance and teaching of employment skills.

In an information-based organization, *knowledge and understanding of complex issues* and their interrelationship with various factors and disciplines *will become a necessity for success*. Specialized knowledge is sterile if it is not put into context. More rapid innovation and its acquisition require understanding of the opportunities, challenges, and potential impact upon the environment and society at large.

Research indicates that many individuals are willing to advance their training skills by attending numerous seminars and workshops, even on their own time. However, employees must also acquire knowledge that may not seem immediately applicable to their jobs but that expands their horizons and helps them to be better citizens. Most universities have not yet offered such educational opportunities to non-degree-seeking students because they still concentrate on teaching various skills through continuing education courses. This conflict must be solved to arrive at an educational system that advances both the individual's knowledge and the skills for work.

Governments

The most important factor for governments is to create the environment and economic climate that attracts industry to locate within the country's borders. Hence, activities that support innovation efforts and technology acquisition by firms are only one small part of the equation and have too narrow a focus. Educational facilities, policy, and economic and political stability are at least as important and are discussed here.

Governments are clearly the most important supporter of education (facilities and funds). As Levin and Schuetze (1983) have pointed out, most government efforts in education are spent providing individuals with basic education. Unfortunately, too often governmental decisions are oriented towards short-term political criteria, such as immediate popularity, but may prove undesirable over the long term.

Education has mostly been concentrated on the younger portion of a country's population, usually ignoring midlife taxpayers. There have been some notable exceptions such as special training programs offered for people in midcareer in Sweden, and retraining opportunities in West Germany and Holland for older employees (see Levin & Schuetze, 1983). Governments must question whether stacking most of one's education into the early part of life is beneficial. Longitudinal research suggests that intellectual flexibility is highest right after leaving school and that it decreases over the years (Kohn et al., 1983). Intellectual flexibility is required, however, to cope with change in one's workplace. Spending six months to a year away from work for more training and education may, therefore, again help many employees to increase their intellectual flexibility and somewhat expand their slightly narrowing perspective. This in turn would facilitate their acceptance of change due to new technology. An interesting option might be to allow each employee to take a six-month educational sabbatical after 10 years of full-time employment. During the sabbatical, the individual would be paid some percentage of his or her average salary for the last 10 years. Financing of such a recurrent education system could be shared equally by the firm and the employee, and the employee would agree to return to the firm for at least one year following the leave.

Another important challenge for government and policymakers is how government agencies and state-owned companies will handle technology acquisition. Governments themselves must assure that they use technology effectively to sustain competitiveness. Ineffective use of technology may increase the price for an essential service. For instance, both the U.S. and the Canadian post office organizations have not yet managed to use computerized mail sorting as efficiently and economically as private companies such as United Parcel Service (UPS) and Purolator Courier.

Even more important is how government-owned firms deal with ILMD. As in the recent Air Canada strike (preholiday season in 1987), the government, through its arbitrator, allowed the union to win by giving its members a contract including automatic cost-of-living adjustments for pensions. The firm, however, might have difficulty meeting these commitments if the government decides to raise the inflation rate for whatever reason. Worse is the fact that unions will push for the same deal in private firms, citing the Air Canada case. As a result, international competitiveness is again lowered. Clearly, industrial policy by the government must begin with how it deals with ILMD in its own firms and agencies. Political short-sightedness could be costly in the long term, especially in the above case as Air Canada was privatized in 1988-1989. Consequently, the airline must cope with a severe long-term handicap while competing in a global market.

Political stability is another factor to be considered. Volatile changes in policies make the management of organizations more difficult and the accomplishment of effectiveness a near impossibility if the parameters change considerably. For instance, the Dutch Prime Minister Ruud Lubbers wrote into the Queen's 1988 Throne Speech that the state of the environment had improved to a satisfactory level. Yet during his campaign for reelection in 1989, he claimed that the 1990s would require drastic measures to further protect the environment. Such changes in political outlook make it difficult for business to plan and respond appropriately over the long term.

To further illustrate this point, during an international conference for environmental protection in Toronto in 1988, Canada agreed to reduce its carbon dioxide emissions by 20% by the year 2005. In September 1989, a conference held by the Canadian government discussed a report with data indicating that, by investing about $128,000,000 in technologies to help reduce energy consumption and harmful emissions (increase in use of public transportation, high-mileage cars, better insulation of houses, and a shift to alternative energies such as solar), energy savings totaling $99,000,000 could be realized. Most important, however, fiscal measures and new laws must be put in place immediately to realize this laudable goal by 2005. In particular, industry must be encouraged to use new technologies (e.g., solar power and insulation materials) at once, as the required changes in energy consumption will require time for innovation and improvement of current technologies (e.g., solar power and car engines). Unfortunately, government bureaucracy will probably not provide legal and fiscal measures before the mid-1990s, thereby making the accomplishment of the objective by the target date a near impossibility. In contrast, Sweden's government decided in March 1988 to close down the first nuclear reactor by 1995, and, by 2010, all of Sweden's nuclear power plants will be shut down. Hence, business can respond to this challenge by developing alternative technologies over several years.

Economic instability is yet another area of high impact; labour market regulations could become so difficult that companies may prefer to ask employees to work overtime rather than hiring new workers, as is the case in West Germany. The reason for this behaviour is that, once a person has been hired, it is extremely difficult under labour law and union contracts to fire him or her, so it may be safer for the firm to avoid bringing unknown elements into their fold. While Volkswagen asks its workers in Germany to work overtime, its Spanish workers' union presses the company to hire new employees instead. SEAT, Volkswagen's Spanish subsidiary, hires about 12.5% of its total work force on term contracts of three years duration, and the rest of the work force enjoys permanent employment with very high job security, which makes it almost impossible for the firm to lay off or let go any of these employees (Knipper, 1989). The labour laws and regulations provide extensive job security for these employees, but the firm tries to cope with these constraints by finding alternatives that give it some leeway.

The above suggests that governments must consider numerous factors that influence technological effectiveness. Innovation and its acquisition by firms may be a minor factor in overall economic success, unless the government provides a conducive environment. Unfortunately, the current tendency is to believe an illusion, that is, that the future and economic success are found in innovation and its successful and early acquisition in the production process alone.

Hence, the technology acquisition-effectiveness equation is affected by various factors but especially by government decisions that affect government business units — such as post offices, utilities, and others — that may erode the competitiveness of private firms. This is especially true as neither unions nor employees usually take international competitiveness, such as comparative labour costs, into account when trying to deal with technology acquisition and organizational adaptation. The latter may reduce employment but increase productivity by individual employees, which in turn may require the firm to raise compensation and thereby eliminate most of the benefits obtained through technology acquisition. The government's responsibility towards business, employees, and voters is far more complex in the long term than is often assumed by the electorate.

Employees

Offering training and continuous education opportunities is one side of the coin; the other is an employee who is willing and eager to upgrade or refresh needed job skills. The individual usually has to decide between employer-sponsored and other programs. The former has the advantage that most of the costs are paid by the organization, attendance will be during working hours, and specially incurred costs, such as travel, are covered as well. In contrast, self-initiated training usually means that the employee is attending courses on his or her own time and costs are probably not paid by the firm. The organization may particularly hesitate to pay if the training is seen as having no immediate benefit for the employee's current position. An employee who is ready for growth may be misperceived as preparing to seek employment elsewhere in the near future (see BIGA, 1988). Moreover, employees must be aware that firm-initiated training is likely to result in firm-specific skills and focus primarily on procedural knowledge, which are not necessarily transferable (e.g., Shaiken, Herzenberg & Kuhn, 1986). Training provided by educational institutions is most often general and more focused towards declarative knowledge, thus allowing the employee to use the skills in other work settings as well as the current one.

One strategy for employees to maximize employability is, therefore, to continuously refresh or upgrade their skills. Another is to keep informed about their organization and its financial situation and position. For instance, internationalization could have a major impact on workers. The Common Market serves as a prime example today. By 1992, it is supposed to be fully integrated, which means, among other things, that all trade barriers will have been removed. Thus a company facing high labour costs in, say, Denmark could move to Spain or Greece, where total labour costs (wages plus fringe benefits) are only about half.

Technology can help high-cost producers to remain competitive by reducing their labour costs. Nonetheless, if the employee's demands for working-hour reductions and salary increases beyond cost-of-living adjustments are not accompanied by increased efficiency and effectiveness in the technology domain, international competitiveness is lowered. North American workers are now coping with the Free Trade Agreement between Canada and the United States, on the one side, and the more limited one currently being discussed between Mexico and the United States. These accords will allow companies to shift production wherever labour and other costs are lowest, thereby improving profitability. Clearly, employees must become aware of the international scope and considerations affecting their employability, for without doing so they may price themselves out of the market.

Unions

Up to now, unions on both sides of the Atlantic have been primarily concerned with job security. The service industry unions, particularly those representing office employees, are now also dealing with compensation aspects (salary, fringe benefits, vacation, and working hours). British research shows that long-term unemployment (6-24 months) negatively affects one's health as well as one's financial situation and that reemployment leads to health improvements (Payne & Jones, 1987a). Furthermore, those whose health was the worst when unemployed reported that job security was most important to them (Payne & Jones, 1987b).

Because continuous employment is important to one's well being and financial health, it is understandable that unions emphasize job security. Nevertheless, job security with increased fringe benefits may endanger the firm's competitiveness. Some organizations in industrialized countries have already started to move some of their office work offshore, with more and more jobs being lost to countries with lower labour costs. Unemployment rates of approximately 6% have become socially acceptable (e.g., in West Germany) and are even declared as representing full employment (e.g., in the United States). From a social and individual point of view, the crucial question is this: What is the lesser of two evils, freezing wages and fringe benefits for a couple of years or losing one's job?

Unions in industrialized countries are faced with a difficult choice. Organizations will try to mechanize certain office jobs even further as machines need neither coffee breaks nor vacations. Accelerating labour costs also encourage such behaviour. In a recent Canadian case, a union had the option of either seeing a plant reduce employment by a third and increase production substantially without pay increases till profitability was once again achieved or having the plant moved and restarted without a union. Social responsibility towards its members and the community left the union no choice but to accept reduced employment.

Union leaders must also increase their knowledge about international trade and comparative wages. For instance, as long as the highest union official within SEAT incorrectly believes that German colleagues at Volkswagen headquarters earn three times as much as the employees of its Spanish subsidiary, Spanish union members' expectations may be raised to a level that is impossible to satisfy (Knipper, 1989). Thus future contract negotiations are destined to end in deadlock. In fact, their incomes, when compared with their different costs of living, are reasonably close. The above would suggest to union leaders that they need to better inform their members about the international perspective when it comes to renegotiating contracts (see also Part III of this book). Additionally, unions, with notable exceptions as the Swedish case suggests (see Rubenson, 1983), have not sufficiently encouraged their members to upgrade their skills. Moreover, the adversary stances taken by management and employees as propagated by many unions often make the union representative look good, but they tend to blur the fact that workers' jobs are at risk while the union representative remains employed. Better information and less adversarial behaviour may eventually be of benefit to all, assuring improved effectiveness in technology acquisition.

Managers

For managers, the major challenge is to understand how technology can be used to promote organizational effectiveness. Clearly, more extensive use of technology in office settings provides managers with new opportunities and options to act as catalysts in organizational change. Organizations must, therefore, learn how to deal with technical, strategic, environmental, and, especially, human resource issues as outlined throughout this book. The major point that bears repeating here is that managers must decide how any technology should be adopted in an organization. More important yet is the manager's understanding and integration of organizational structure and processes, ILMD, and choice in SHRM in the implementation process. Ignoring either may have serious consequences for the firm. The necessary human resources for acquisition and implementation of technology are affected by the firm's HRM strategy and various human resource systems, as discussed in Part II of this book.

A major challenge for managers is to modify their traditional approaches to management. Natural open systems evolve and adapt based on external and internal opportunities, challenges, and threats. As the above shows, before technology acquisition is accepted, unions as well as employees may demand certain concessions that could threaten organizational competitiveness and even survival. Most companies have failed to address this issue strategically and with consideration of their human resources. Instead, many adopt technology in a rather unplanned and sometimes piecemeal fashion. This course may not result in immediate failure, but resistance by the work force towards technology-induced change could develop due to a feeling that job security is being threatened or that employees have been misinformed. Another difficulty is that managers themselves often avoid adapting their own work habits to technology. For instance, even though computers may be at their workplace, managers tend only to use them for peripheral job activities (Floyd, 1988). Effectiveness in EUTM starts, however, at the manager's own desk. If he or she decides to rely on an assistant's computer literacy, effectiveness is difficult to achieve. More important, however, managers are role models who are often copied by subordinates. If it becomes a power symbol to avoid using new technology, once again, employee resistance is certain to occur in various parts of the organization.

Another important challenge is better communication (formal and informal) with employees by managers. Financial and strategic information should be passed on to the employees via meetings and/or newsletters. A strong information flow would improve the relationship between management, employees, and the union, making the latter's antagonistic stance unnecessary.

Although participation in an executive committee and in the formulation of the firm's technology strategy by employees is most often not legally required, its implementation by management will again increase the dissemination of information and improve its accuracy among the different levels within the organization. Rapid technology acquisition and the subsequent continuous change within the organization require that adequate information be held by all parties. This is an opportunity for management that should not be missed.

Remarks

Much of what has been written about technology acquisition has been speculative and some of it has been ill informed. Most of the early forecasts about the revolutionary impact of the microchip and computers have turned out to be grossly exaggerated. And there are as yet relatively few empirical studies of technology acquisition upon which to build well-founded generalizations.

What I have attempted to do in this book is present a representative collection of available research evidence about technology acquisition within firms as new technology permeates the workplace, and to discuss these findings in the light of our theoretical knowledge about the behaviour of firms, managers, unions, and employees. Most important, however, is the attempt to develop an integrative framework to help researchers and others refocus their own work in this area, thereby increasing our knowledge about the *process* of technology acquisition.

The research approaches discussed and the models developed in this book illustrate the fact that technology acquisition should become a central and explicit variable of long-term global, social, and corporate strategy and policy, and not just the instrument of short-run national or organizational gain witnessed as many governments and firms today jump headlong onto the high-technology bandwagon. Of foremost importance is a better understanding of the mechanisms occurring during technology acquisition, which can be gained by putting the human being at the core of this process. Participation by the work force at all stages of the acquisition process helps the organization ensure better communication among all parties involved, consequently improving understanding of the various contingencies (environmental, ILMD, and individual) the firm must deal with to succeed. This will help in smoothing the organizational adaptation process and, consequently, increase both monetary and nonmonetary benefits from new technology.

What, within this short and partial account, are the lessons? One obvious answer is, of course, that, if negative effects of technology acquisition are to be confronted, shifting emphasis from the quantitative towards the qualitative domain of acquisition and organizational adaptation assessment must be accomplished. Is a path of technology acquisition that focuses mostly upon financial data and environmental determinism feasible, when much evidence suggests that acquisition is a complex process of serendipity? Such a rational approach may not discover why and how personal values, feelings, and interests may have led some participants to decide about technology acquisition in a manner that appears irrational from an objective outsider's perspective. Furthermore, should not the internationalization of business and the cultural diversity within most nations be taken into account by also assessing cross- and intracultural differences and their effect upon technology acquisition?

Organizations should address all of the above issues to become truly successful in their technology acquisition and organizational adaptation efforts with any type of technology. Moreover, the employee is the crucial link between all the interrelating dimensions, as outlined in Figure 13.1. This book represents a first step in the right direction, but much additional work must be done before technological innovation can be managed in such a fashion as to increase the derived benefits for both the firm and the employee. Additionally, the latter must be assured that quality of work life will not diminish due to technology acquisition. Work is an important part of a person's life, often providing its main purpose; degradation of work is, therefore, highly undesirable and should be avoided at all cost.

As Pfeffer (1982) and others have argued, the gospel that has been preached is that proactive and farsighted individuals make choices based on rational analyses of the situation. Unfortunately, most researchers and practitioners forget that values and ideals permeate any theory about technology acquisition. Hence, technology acquisition based on the erroneous assumption of rationality should be reassessed. Results obtained have been meager, and other perspectives, some of which have been discussed in this book, have at least the potential for offering parsimonious and more valid analyses of technology acquisition. Technology acquisition and new developments will continue to force organizations to make better use of opportunities to remain competitive in national and international markets. The fundamental purpose of this book has been to help those taking part in the technology acquisition process—whether researchers, employees, managers, shareholders, or others—better understand the limitations hindering them and the possibilities open to them and their firms when it comes to successful technology management.

Glossary

Acceptance/Resistance: Accepting or resisting technology is governed by (a) *determinism*, which states that attitudes are not a function of deep-seated needs but a product of social information; and (b) *social comparison*, where the individual specifically compares his or her work situation with others.

Acquisition of Technology: This is the organizational process that includes planning, selection, financing, purchasing, or leasing and the subsequent introduction of a specific technology in the workplace.

Action Strategy: This is one's way of organizing action.

Amadeus: A European computerized reservation system, similar to *SABRE* in North America and *Galileo*, also used in Europe, that is being implemented in response to competition from American airlines trying to establish themselves in European markets. Participating airlines include Air France, Iberia, Lufthansa, and SAS.

Anticipatory Socialization: This encompasses all the learning that occurs before a new member joins an organization. Such learning might come from talking to current employees or reading about the firm or from information provided about the company before a job interview. Assuming that this information gathering activity provides an encouraging view of the firm and the job, he or she might then begin to unconsciously accept the values of the organization – in essence, to socialize him- or herself.

Appraisal: This is the process and product of evaluating an employee's performance after the initial phases of *socialization*.

Authority Structures: These may consist of a clearly defined hierarchy with formal boundaries, based on seniority (such as in *Quadrants I and IV*), or a looser structure based on influence and expertise (as in *Quadrants II and III*).

Automated Production: Mechanized production is made "automatic" by using the feedback-control principle, thus reducing the degree of human decision making necessary to keep the production process going.

Basic Skills: These skills consist of reading, writing, and arithmetic.

Behaviourally Anchored Rating Scales (BARS): This is a graphic rating scale with specific behavioural descriptions using various points along each scale. Each scale represents a dimension or factor considered important for work performance. This is a method of *appraisal*.

Bureaucracy: This includes any open system whose administrative and production processes have been slowed down and are extremely cumbersome.

Career: Traditionally, this has been defined as a linear progression up the corporate ladder.

Career Development: This personal development is also viewed in relation to career choice and entry as well as progress in educational and vocational pursuits.

Career Development System: Used to coordinate *career development* activities – an organized, planned, ongoing effort to achieve a balance between the individual's career needs and the organization's work force needs – it integrates the activities of employees with the policies and procedures of the firm and facilitates the development and refinement of present human resource activities.

Career Management: Included under the umbrella of *human resource management*, this encompasses career planning and counselling, assisting individuals with self-assessment, making career and job changes possible, and managing both organizational career paths and the internal work force.

Categorization: This theory states that individuals perceive and process information in terms of abstract categories defined either by various schemata or, more rarely, by somewhat more concrete prototypes.

Change and Acquisition Socialization: In this phase of *socialization*, relatively long-lasting changes take place; the employee masters his or her job, adjusts to the work group's norms and values, and is successful in performing his or her new role.

Communication System: Evolving out of the *authority structure* and *negotiation strategy* in a firm, this system may be vertical, strictly between employees and immediate supervisors and backed up by numerous regulations and conventions, as in *Quadrant I*; vertical with more freedom for employees to communicate with upper management, as in *Quadrant II*; governed by guidelines that permit communication across all levels, as in *Quadrant III*; or unlimited but unexploited, as in *Quadrant IV*.

Compensation Systems: These may consist of a firm's *remuneration system*, *reward system*, or other individualized compensation packages such as cafeteria-style benefits, unpaid leave, or early retirement. Prestige, job security, and other social (as opposed to financial) benefits may also be included in such a system.

Computer-Aided Manufacturing: As the latest step in the evolution of mechanization, this allows production plants to run with a minimum number of workers according to preprogrammed instructions.

Conceptual Skills: These skills include planning, assessing, judging, and decision making about tasks and people-related issues.

Critical Incident Method: This method of *appraisal* consists of collecting reports of behaviours that are considered "critical," in the sense that they make a difference to an employee's success or failure in a particular work situation, and assessing performance based on these.

Cultural Repertoire: Individual habits, skills, and values as they relate to *culture*.

Culture: Comprising people's attitudes, beliefs, and rituals (see Chapter 10 for further definitions), culture has been described as a rather elusive construct without clearly drawn boundaries.

Culture-Free Hypothesis: This hypothesis argues that the relationship between certain contextual variables and dimensions of organizational structure are similar across very different societies.

Decision Making and Negotiation Strategies: Varying according to organization type, in *Quadrant I*, decision making and negotiations are centralized and strictly regulated; in *Quadrants II and III*, decision making is decentralized and the negotiation strategy may be flexible or guided by clear policies depending on environmental conditions; in *Quadrant IV*, decision making is centralized but no definite policy for negotiation exists.

Declarative Knowledge: This is knowledge *about* something.

Derivative Innovation: Utilization of *primary innovations* in applied research that is based in part on marketplace demand or other environmental factors such as government regulations.

Glossary

Deskilling: Management attempts to transfer control of work to management by depriving employee of the use of his or her skill.

Differentiation Strategy: A strategy aimed at creating a product or service that is somehow different.

Diffusion: This is the degree to which an innovation has become integrated into an economy. Five factors influence the rate at which a technology diffuses: the origin of the innovation, the effects on other inputs, the relationship of the innovation to the existing production structure, change in the innovation, and complementaries among innovations.

Discontinuous Introduction of Innovation: A process by which innovation is introduced in a rather short time frame, the change in technology in this sort of innovation would be radical.

Distinctive Competence: Used to describe the character of an organization, this refers to those things that a company does especially well when compared with others in a similar environment.

Doctrine of Equal or Comparable Worth: Employees holding different jobs judged to be of similar value will be paid the same regardless of gender and the external market value of their skills.

Encounter Socialization: This begins with the employee's first day on the job; from this firsthand experience of the organization, initial impressions gained from *anticipatory socialization* may change.

End-User: These individuals work with technology but have a limited amount of technical skills.

Environmental Determinism: People believe that there are factors in the environment that cannot be chosen or changed but must be accepted the way they are; for example, certain characteristics of industries and national economies are intractable, immune to the control of individuals or their organizations.

Evolutionary Introduction of Innovation: In this process, innovation is introduced gradually, step by step, allowing each step to evolve from the preceding one.

External Labour Market: As the term *external* implies, this labour market is outside an individual organization, consisting of employees of other organizations who may wish to change jobs, graduates of training programs, and those dismissed from former positions.

Flexible Work Scheduling: Employers may contract employees to work a certain number of hours per week without stipulating the number of hours they will work on any given day; employees are then required to work overtime on days when the demand for their skills is great, without receiving overtime rates.

Forced Distribution Method: This method of *appraisal* requires the appraiser to "force" a designated portion of the subjects into one of several categories of performance.

Governance: These structures exercise authority over an organization and interact or influence management. In general, profit organizations are controlled by owners who may also be shareholders. Management is assisted by a board of directors, which may include voting and nonvoting members who could represent such diverse groups as employees, management, unions, customers, and suppliers.

Graphology: This analysis involves the creation of a freestyle overall qualitative personality description based on a handwriting sample submitted by a job applicant.

High Technology: An as yet undefined concept including the production of advanced machines, design and software, and organisms (e.g., genetic engineering), it is pertinent to manufacturing, laboratories, and offices.

Human Capital: This is the total accumulation of the skills, knowledge, and talents present in a firm's work force.

Human Resource Culture: This summarizes the managerial approaches and attitudes towards personnel, explaining the state of the *organizational culture*. Strategies for maintaining a company's cultural identity in terms of prevailing values may or may not exist but tend to contribute to organizational effectiveness if present.

Human Resource Management (HRM): This is the attraction, selection, retention, and utilization of human resources to achieve an effective work force and the attendant individual and organizational objectives.

Human Resource Management Strategies: These strategies are formed from the decisions and choices made over time by top management concerning human resources.

Ideal Candidate: This theoretical individual would best fit organizational requirements during selection of new personnel.

Incremental Innovation: Management introduces technology new to the organization that is very similar to its present technology, which requires little change in the processes of the organization in order to be used effectively.

Innovation: A process or the product of a process that is the result of the efforts or activities of an individual, group, and/or organizational system, that represents a departure from the previous state and may facilitate more effective resource allocation. *Innovation* is considered synonymous with *technological innovation* in this text.

Institutional Environment: This environment includes the rules and belief systems as well as the relational networks that arise in the broader societal context.

Insufficient Justification Hypothesis: These effects of wage inequality are *not* cancelled out by job interest and/or lack of financial need.

Internal Labour Market (ILM): The set of administrative rules and procedures that govern the pricing and allocation of human resources and training decisions within an organization, its allocative dimension is its essential feature, which is used to allot human resources to positions within the organization based on rules and regulations, necessitating internal promotion and limiting the number of ports of entry into the organization.

Internal Labour Market Culture (ILM Culture): This comprises the basic assumptions and beliefs shared by members of the firm concerning the bureaucratic rules and regulations about work, job content/classification, and entry/exit barriers.

Internal Labour Market Determinism: These forces limit the adoption of innovation or new technology, including rules, rigid bureaucratization of staffing and rewarding, and union contracts that can be altered only after a lengthy bargaining period, leading to difficulties in changing job duties and skill requirements.

Job Analysis: A possible facet of *selection*, *job analysis* is a procedure that captures the behaviours exhibited and activities performed by the present job holder.

Job Classification System: The system is used to cluster jobs for wage and salary administration on the basis of common skills, qualifications, technology, and working conditions.

Job Content Compensation: This compensation is in the form of improved working conditions; for example, increasingly challenging tasks or more influence on company policy, status, authority.

Job Sharing: Two or more people share the same job; sometimes called "work sharing."

Labour Relations: This encompasses the negotiation and administration of union contracts.

Lateral Mobility: This interorganizational transferability is dependent upon the individual's willingness to perform different tasks and acquire new skills required in his or her work.

Glossary

Learning: In an organizational context, *learning* may be thought of as a process by which an individual's pattern of behaviour is altered in a direction that contributes to organizational effectiveness.

Macro: This describes analysis at the organizational level.

Mechanization: As a growing reliance upon machine power, as opposed to human effort, this process began in the late eighteenth and early nineteenth centuries and continues to this day.

Mentors: New staff can turn to specific senior employees to learn the ropes.

Merit Pay System: This *compensation system* differentially rewards employees on the basis of relative and (occasionally) absolute assessments of their performance. Performance criteria may or may not be defined; such systems often allow for nonperformance factors to be taken into consideration.

Meta-Analysis: As a family of techniques, it is used to aggregate research evidence across studies.

Micro: This describes analysis at the individual level.

Mobility: One moves up in the corporate hierarchy, as measured by the frequency of promotions.

Multidivisional Organizational Structure (M-Form): Organizing a firm's activities using quasi-autonomous operation divisions, divided mainly along product and geographical areas (decentralized control), this structure is used most frequently by larger firms, where the dispersion of operation responsibilites among division managers leaves the corporate management group free to focus its energies on problems of overall strategic direction and entrepreneurial action. This type of organization structure may facilitate the adaptation of innovation.

Negotiation: A process of discussion between individuals or groups with different interests with a view to reaching agreement about goals, which is aimed to result in maximum *organizational effectiveness*.

Objective Culture: This comprises the ecological system, subsistence system, and parts of the cultural (manmade part of the environment) system.

Open Systems: These systems are open to external influences and dependent on a flow of resources and personnel from outside; a coalition of individuals with various, constantly shifting interests develops and agrees upon goals by *negotiation*.

Organizational Choice: One is able to choose in decision-making circumstances and construct, eliminate, or redefine the objective features of the environment to expand decision-making capabilities.

Organizational Culture: Consisting of the basic assumptions and beliefs that are shared by all members of the same company, these beliefs usually operate unconsciously but define how the organization sees itself and its environment and may be reinforced by a reward system.

Organizational Effectiveness: This describes the fulfillment of the goals of the organization's dominant coalition of stakeholders and may be dependent upon the successful resolution of conflicts between participants through *negotiation*.

Organizational Structure: A configuration of the decision-making and information processing activities (i.e., the daily, weekly and quarterly cycle of routines, procedures, reports, and forms) that tends to be enduring and persistent. Patterned regularity is the dominant feature of this configuration. The two main types of *organizational structure* are *unitary* (U-form) and *multidivisional* (M-form).

Organization of Work: Varying according to organization type, in *Quadrants I* and *IV*, which are often *U-form* organizations, jobs are more likely to be defined rigidly according to handbooks and hierarchic patterns, while in *Quadrants II* and *III*, which are likely to be organized according to the *M-form*, jobs cannot be as clearly defined.

Ownership: One owns part of an organization through company stock and can influence managerial tactics; executives with limited ownership are more likely to be concerned with reducing overhead than with adopting innovations that may be expensive initially, an attitude likely to gain stakeholder's favour.

Paired-Comparisons: A method of *appraisal*, it requires the appraiser to make relative comparisons between subject employees in terms of performance or organizational worth. The supervisor may compare all possible pairs of subordinates on their overall ability to do the job.

Pay: This compensation is in any form: wages, fringe benefits, cash bonuses, and so on.

Pay Compensation: This monetary compensation is in the form of wages, bonuses, and/or fringe benefits.

Pay Levels: As rates of *pay* for different positions in the firm, these will vary both between and within organizations.

Pay Structures: These are defined by the relationship of *pay levels* across occupations, jobs, grades, and other classifications within the firm.

Pay System: This system comprises the set of contingencies that determine *pay levels* and changes in these levels.

Performance Appraisal: Involving the development of criteria based on either the behaviour of others or a standard established by management, this process reviews an individual's performance (the appropriateness or desirability of the employee's behaviour within the organizational setting) and develops subsequent performance objectives. It is designed to give employees the feedback they need to maintain or improve the quality of their work.

Personality Testing: A possible facet of *selection*, *personality testing* is the measurement of personality traits using psychological testing devices.

Population Ecology/Natural Selection Approach to Adaptation: This approach argues that organizations enjoy virtually no control over exogenous factors. Instead, adaptation is in part determined from outside the firm, as the environment selects the firm and allows only those with appropriate variations and products to remain profitable. Under these conditions, firms must either adapt technology according to external determinism or cease to exist.

Primary Innovation: Accomplished by primary research, it is based on work evolving out of the testing of theoretical relationships. For example, the invention of plastic was a *primary innovation*.

Proactive Change: This describes change adopted when it is untried and innovative; an organization that is *proactive* embraces technological innovation likely to give it a competitive edge in the marketplace before competitors do so.

Procedural Knowledge: This is knowledge about *how to do* something.

Processes: Human resource strategies are developed to facilitate interaction between *organizational structure*, *human resource management*, and the *internal labour market*.

Product and Process Innovation: Traditionally, *product* innovations represent radical changes from past systems, and *process* innovations represent incremental changes, with improvements leading to a rigid, efficient production system specifically designed to produce a standardized product. Recent research has suggested that a binary distinction cannot capture the reality of innovation and change and that these two types of innovation

Glossary

are on a continuum. This approach is reflected in this text, which does not distinguish between *product and process innovation.*

Profit-Sharing: In this *compensation system*, workers accept reduced fixed hourly wages and benefits in return for a share of the organization's profits.

Promotion Ladders: These clearly and formally defined pathways provide promotion opportunities within a firm.

Prospectors: These individuals or companies are looking for the next "gold rush" in the form of an innovative new product that will mean overnight success.

Public Ownership: The shares of a company and those of its subsidiaries are listed on several stock exchanges.

Quadrant I Firms: Marked by an absence of organizational control over exogenous factors, these firms are usually found in markets that are oligopolistic or highly regulated. Generally, these organizations exercise little control because the market dictates the price for their product while the demand for it is fairly elastic. Technological adaptation is often due to environmental determinism.

Quadrant II Firms: These firms are marked by high levels of strategic choice and environmental determinism. Certain rules, constraints, or immutable environmental conditions compel some outcomes or behaviours but allow leeway and choice in others. Technological adaptation is less restricted because companies and industries will be producing a variety of products and services.

Quadrant III Firms: Marked by high levels of organizational choice, autonomy, and control, the organization in Quadrant III confronts a pluralistic environment in which movement between niches of markets is not severely constrained by exit or entry barriers. Further, companies in this quadrant neither lack scarce resources nor experience many political constraints and are, therefore, able to define their domains and the exogenous conditions under which they desire to compete. In Quadrant III, technological adaptation is by design.

Quadrant IV Firms: Firms in this quadrant are in a relatively stable situation yet tend to lack strategic choice, despite a paucity of external constraints. Changes in technology seem to occur almost by chance, because these firms seem to exhibit no coherent strategy to take advantage of fortuitous market and product conditions.

Radical Innovation: This describes the introduction of technology that is very different from that which the organization has previously used, which, therefore, requires significant changes in processes if it is to be used effectively.

Rank Ordering: In this method of *appraisal*, the appraiser first selects the "best" person, then the "worst" person, then the second best, then the second worst, and so forth until all persons have been ranked.

Reactive Change: This kind of change is adopted in reaction to a threat that is the result of environmental changes. An organization that is *reactive* only adopts new technology if threatened by the fact that a competitor is already using it.

Recruitment: As the activities or practices that determine the characteristics of applicants to whom *selection* procedures are ultimately applied, recruitment occurs prior to *selection* and determines the type of applications from which predictor information is gathered for use in the final hiring process.

Recurrent Training: Training is updated throughout an employee's career.

Reliability: This describes the extent to which an appraisal method/instrument provides consistent results.

Remuneration System: This is part of the overall compensation package offered by the firm to both salary and wage employees; the pay and fringe benefits that an organization is legally required to provide to employees.

Resource-Dependence Model: This model suggests that individuals whose performance is more important to the organization's well-being will be treated differently than less valuable employees.

Reward System: As a subsidiary compensation system, it provides incentives that direct and motivate employee behaviour towards improving individual and collaborative performance. Established criteria often must be met to qualify for these incentives.

Selection: This procedure involves evaluating predictor information for the purpose of making a final hiring decision.

Sequential Work Flow: The worker performs a specific job on basically the same part, component, information, or other throughput and then passes the part on to the next workstation (assembly line).

Skill Specificity: This is firm-specific training given to employees as well as the degree of transferability of firm-specific training to other employers.

Social Comparison Theory: Individuals selectively use comparison information to learn about themselves *and* the social environment.

Social Information Processing Model: This model states that perceptions of and approaches towards tasks are at least partially a function of social cues in workplaces, which implies that perceptions of job content are based upon the individual's interpretation of events or situations within their work contexts.

Socialization: This describes a systematic attempt to imbue the employee with the values and norms that characterize the organization.

Socialization Systems: These systems are designed to acclimatize employees to the organizational environment — to help them "learn the ropes" and become comfortable within the organization.

Social Skills: These skills include the ability of employees to organize their own effort, individually or as part of a team, and to get along within a work group. For managers, they include organizing their peers and subordinates.

Specialization: Employees acquire different skills needed to perform only work activities attached to a specific position and/or job category.

Stage Strategies for Adopting Technology: There are five different stage strategies: *differentiation* strategy, which aims at creating a product or service that is somehow different; *cost leadership* strategy, which aims at making the organization the lowest-cost producer in an industry; *focuser* strategy, which caters to a circumscribed and specialized segment of the market (e.g., a certain kind of customer, a limited geographic market, and/or a narrow range of products); *reactor* strategy, which is used to respond to the actions of competitors; and *defender* strategy, in which the firm tries to defend its market share, typically by lowering prices or by dumping.

Stakeholders: This describes everyone with an interest in the activities of an organization (e.g., shareholders, employees, customers, and members of the general public who are affected by the company's endeavours).

Status Differentiation Compensation: Compensation is in the form of increased status for key employees; status indicators may include the type of office and/or furniture, travel arrangements, or even a company car.

Glossary

Status Quo Strategy: This recruiting strategy emphasizes individuals who have been recommended or referred to the firm by other employees and by community leaders in particular.

Strategic Choice: One has ability to choose in decision-making circumstances and construct, eliminate, or redefine the objective features of the environment to expand decision-making capabitities.

Strategic Human Resource Management (SHRM): This describes the attraction, selection, retention, and utilization of human resources to achieve an effective work force and the attendant individual and organizational objectives over a relatively long period (from three to five or more years).

Strategic Management: This decision-making process focuses the capabilities of the firm upon the opportunities and threats it faces in its environment. The purpose is to create and maintain systems that enable employees to react appropriately in any given situation and that facilitate organized action by the firm.

Strategy: See *strategic management*.

Structural Contingency Theory: This theory is based on a "law of interaction," which states that the performance of an organization is dependent upon a fit between the firm's *structure* and *processes*, given that normal assumptions hold about the premises, boundaries, and system states derived from the theory. Good fit between structure and technology increases effectiveness.

Subjective Culture: This comprises the social system, the cultural system (norms, roles, and values as they exist in the individual's environment), and individual perceptions.

Systematic Risk: This describes overall market risk which cannot be diversified away.

Talent Additive Activity: This describes an activity in which each individual's output contributes more or less equally to total organizational output; disproportionate contributions are not possible.

Talent Complementary Activity: This describes an activity in which an individual's performance can add disproportionately to total organizational output.

Task Identity: This describes the degree to which a job requires completion of a whole and *identifiable* piece of work.

Task Skills: These usually task-specific skills are required to perform a job-related task.

Technical Skills: These skills consist of the physical ability to transform an object or item of information into something different.

Technological Adaptation: In this text, it may be used to refer to the process of dealing with both *reactive* and *proactive change*. Organizational factors like structure, corporate strategy, and environmental determinism are related to *technological adaptation*.

Technological Innovation: This technology-based process, or the product of such a process, results from the efforts or activities of an individual, group, and/or organizational system that represent a departure from the previous state and that may facilitate more effective resource allocation. This term is considered synonymous with *innovation* in this text.

Technology Skills: These consist of the appropriate use of technology in job related tasks.

Tenure: This describes seniority rights embedded in a system of rigid job classification.

Term Contracts: These employment contracts are for relatively short periods of time, allowing employers to pay only legally required benefits such as holiday pay and unemployment insurance. An organization has the choice of whether or not to renew the contract when it expires, thus making *term contracts* an effective method of control over the *ILM*.

Training: In an organizational context, training involves any organizationally initiated procedures intended to foster learning among organization members.

Trait Approach: This is a popular method of *appraisal*, which evaluates employees on the extent to which each possesses such traits as punctuality, dependability, friendliness, and ambition.

Two-Tier Wage Structure: In this *pay structure*, the top level of pay for new employees is substantially lower than that for long-time employees, resulting in personnel with the same job titles and duties receiving different pay for similar inputs of effort. Usually, newly hired employees can eventually work their way up to the higher wage scale of veteran employees.

Unionization: The process of becoming unionized, it occurs when employees join a union.

Unitary Organizational Structure (U-Form): This type of structure organizes a firm's activities along functional specialties such as sales and finance (centralized control); it is used most frequently by small firms, due to the need for careful coordination of strategic direction and control for all departments in the organization. This organization structure may slow the adoption of innovation.

Unsystematic Risk: This risk is unique to an individual firm and associated with the organization's specific resources, competencies, and relationship to the environment.

Validity: This describes the extent to which an appraisal method/instrument measures what it is designed to measure.

Wage Discrimination: This salary gap in organizations is due *directly* to differences in compensation given to human capital based on factors like organizational tenure and education.

Wage Segregation: This salary gap is *indirectly* created by inequalities in the distribution of men and women across job classes and hierarchical ranks.

Work Flow: This describes the organization of the transformation process through which inputs, such as capital, raw material, and human effort, are turned into outputs.

Work-flow Parallelism: The same jobs are done parallel to each other (e.g., several work groups may install various parts of a car simultaneously using parallel workstations).

Work Role: Defined by the behaviours that are characteristic of the role, this describes behaviours that are expected and required of an employee by his or her superiors and the organization.

Work Sharing: Two or more people share the same job; sometimes called "job sharing."

References

Abernathy, W., & Utterback, J. (1978, June/July). Patterns of industrial innovation. *Technology Review*, pp. 40-47.

Abrahamson, M. (1973). Talent complementarity and organizational stratification. *Administrative Science Quarterly, 18*, 186-193.

Ackerman, P. L. (1987). Individual differences in skill learning: An integration of psychometric and information processing perspectives. *Psychological Bulletin, 102*, 3-27.

Ackerman, P. L. (1988). Individual differences and skill acquisition. In P. L. Ackerman, R. J. Sternberg, & R. Glaser (Eds.), *Learning and individual differences* (pp. 123-150). San Francisco: Freeman.

Adams, J. A. (1987). Historical review and appraisal of research on the learning, retention, and transfer of human motor skills. *Psychological Bulletin, 101*, 41-74.

Adams, J. S. (1963). Toward an understanding of inequity. *Journal of Abnormal and Social Psychology, 67*, 422-433.

Andrews, K. R. (1971). *The concept of corporate strategy*. Homewood, IL: Irwin.

Ansen, D. (1989, July 31). Boffo box office big boost to biz. *Newsweek*, pp. 60-62.

Arnold, H. J. (1985). Task performance, perceived competence, and attributed causes of performance as determinants of intrinsic motivation. *Academy of Management Journal, 28*, 876-888.

Arnold, J. D., Rauschenberger, J. M., Soubel, W. G., & Guion, R. M. (1982). Validation and utility of a strength test for selecting steelworkers. *Journal of Applied Psychology, 67*, 588-604.

Arnould, R. J. (1985). Agency costs in banking firms: An analysis of expense preference behaviour. *Journal of Economics and Business, 37*, 103-112.

Ash, R. A., & Levine, E. L. (1985). Job applicant training and work experience evaluation: An empirical comparison of four methods. *Journal of Applied Psychology, 70*, 572-576.

Astley, W. G., & Van de Ven, A. H. (1983). Central perspectives and debates in organization theory. *Administrative Science Quarterly, 28*, 245-273.

Attewell, P. (1987a). Big brother and the sweatshop: Computer surveillance in the automated office. *Sociological Theory, 5*, 87-99.

Attewell, P. (1987b). The deskilling controversy. *Work & Occupations, 14*, 323-346.

Bandura, S. (1977). *Social learning theory*. Englewood Cliffs, NJ: Prentice-Hall.

Becker, G. S. (1964). *Human capital*. New York: National Bureau of Economic Research.

Becker, G. S. (1971). *Economics of discrimination*. Chicago: University of Chicago.

Behling, V. W., Gifford, W. E., & Tolliver, J. M. (1980). Effects of grouping information on decision making under risk. *Decision Sciences, 11*, 272-283.

Belous, R. S. (1987). High technology labor markets: Projections and policy implications. In A. Kleingartner & C. S. Anderson (Eds.), *Human resource management in high technology firms* (pp. 25-45). Lexington, MA: Lexington.

Ben-Shakhar, G., Bar-Hillel, M., Bilu, Y., Ben-Abba, E. & Flug, A. (1986). Can graphology predict occupational success? Two empirical studies and some methodological ruminations. *Journal of Applied Psychology, 71*, 645-653.

Bernardin, H. J., & Beatty, R. W. (1984). *Performance appraisal: Assessing human behavior at work.* Belmont, CA: Wadsworth.

Betriebsrat, Max-Planck-Institut für Bildungsforschung. (1988). *Entwurf einer Betriebsvereinbarung zu individueller Datenverarbeitung im Max-Planck-Institut für Bildungsforschung zu Berlin* [Draft policy statement on individual data processing at the Max-Planck-Institut in Berlin]. Berlin: Author.

Bettis, R. A. (1983). Modern financial theory, corporate strategy and public policy: Three conundrums. *Academy of Management Review, 8*, 406-415.

Bhagat, R. S., & McQuaid, S. J. (1982). Role of subjective culture in organizations: A review and directions for future research. *Journal of Applied Psychology Monograph, 67*, 653-685.

Biddle, B. J. (1979). *Role theory.* New York: Academic Press.

Bielby, W. T., & Baron, J. N. (1986). Men and women at work: Sex segregation and statistical discrimination. *American Journal of Sociology, 91*, 759-799.

BIGA (Bundesamt für Industrie, Gewerbe und Arbeit [Swiss Federal Office for Industry, Trades & Work]). (1988). *Weiterbildung in der Schweiz—Auswertung einer Umfrage* [(Continuous education in Switzerland: Evaluation of a poll)]. Bern: Author.

Bigoness, W. J. (1988). Sex differences in job attribute preferences. *Journal of Organizational Behaviour, 9*, 139-147.

Bikson, T. K., & Gutek, B. A. (1983). *Training in automated offices: An empirical study of design and methods.* (Report No. WD-1904-RC). Santa Monica, CA: Rand Corporation.

Bikson, T. K., Gutek, B. A., & Mankin, D. A. (1987). *Implementing computerized procedures in office settings: Influences and outcomes* (R-3077-NSF/IRIS). Santa Monica, CA: Rand Corporation.

Birkwald, R. (1973). Menschengerechte Arbeitswelt. *Arbeitsheft # 013*. Frankfurt, Federal Republic of Germany: Industrie-gewerkschaft Metall, IGM.

Bjorn-Andersen, N., Eason, K., & Robey, D. (1986). *Managing computer impact: An international study of management and organizations.* Norwood, NJ: Ablex.

Blalock, H. M. (1984). *Basic dilemmas in the social sciences.* Beverly Hills, CA: Sage.

Blau, P. M., & Schoenherr, R. A. (1971). *The structure of organizations.* New York: Basic Books.

Blau, P. M., & Schwartz, J. E. (1984). *Crosscutting social circles: Testing a macrostructural theory of intergroup relations.* Orlando, FL: Academic Press.

Blyton, P. (1987). The working time debate in Western Europe. *Industrial Relations, 26*, 201-207.

Boulet, J., & Lavallée, L. (1984). *The changing economic status of women.* Ottawa: Economic Council of Canada.

Braithwaite, R. B. (1953). *Scientific explanation.* Cambridge: Cambridge University Press.

Braverman, H. (1974). *Labor and monopoly capital.* New York: Monthly Review Press.

Breakwell, G. M., Fife-Schaw, C., Lee, T., & Spencer, J. (1987). Aspirations and attitudes to new technology. *Journal of Occupational Psychology, 60*, 169-172.

Breaugh, J. A. (1981). Relationships between recruiting sources and employee performance, absenteeism, and work attitudes. *Academy of Management Journal, 24*, 142-147.

Brousseau, K. R. (1978). Personality and job experience. *Organizational Behavior and Human Performance, 22*, 235-252.

Brumlop, E., & Juergens, U. (1986). Rationalisation and industrial relations: A case study of Volkswagen. In O. Jacobi, B. Jessop, H. Kastendiek, & M. Regini (Eds.), *Technological change, rationalisation and industrial relations* (pp. 73-94). New York: St. Martin's.

References

Bureau of Labor Statistics. (1982, December). *Employment and earnings*. Washington, DC: Government Printing Office.

Bureau of Labor Statistics. (1984, October). *Employment and earnings*. Washington, DC: Government Printing Office.

Burke, M. J., & Day, R. D. (1986). A cumulative study of the effectiveness of managerial training. *Journal of Applied Psychology, 71*, 232-245.

Burns, T., & Stalker, G. M. (1961). *The management of innovation*. London: Tavistock.

Burton, R. M., & Obel, B. (1980). A computer simulation test of the M-form hypothesis. *Administrative Science Quarterly, 25*, 457-466.

Burton, R. M., & Obel, B. (1984). *Designing efficient organizations: Modelling and experimentation*. Amsterdam: North Holland.

Cameron, K. S., & Whetten, D. A. (1983). *Organizational effectiveness*. New York: Academic Press.

Campbell, A., & Warner, M. (1988). Microelectronics, skill shortages and training strategies: A study of selected British companies in the high-technology sector. *Journal of General Management, 13*(4), 5-32.

Campbell, J. P., Dunnette, M. D., Lawler, E. E., & Weick, K. E., Jr. (1970). *Managerial behavior, performance, and effectiveness*. New York: McGraw-Hill.

Card, S. K., Moran, T. P., & Newell, A. (1984). *The psychology of human computer interaction*. Hillsdale, NJ: Lawrence Erlbaum.

Carroll, S. J., & Schneier, E. C. (1982). *Performance appraisal and review systems: The identification, measurement, and development of performance in organizations*. Glenview, IL: Scott, Foresman.

Carter, N. M. (1990). Managerial attitudes toward computerization: Control, acceptance and employee alienation. In U. E. Gattiker (Ed.), *Technological innovation and human resources: Vol. 2. End-user training*. New York: De Gruyter.

Carter, S. B. (1988). The changing importance of life-time jobs, 1892-1978. *Industrial Relations, 27*, 287-300.

Cascio, C. F., & Awad, E. M. (1981). *Human resources management*. Reston, VA: Reston.

Cascio, W. F. (1987). *Costing human resources: The financial impact of behavior in organizations* (2nd ed.). Boston: Wadsworth.

Casey, B. (1986). The dual apprenticeship system and the recruitment and retention of young persons in West Germany. *British Journal of Industrial Relations, 24*, 64-81.

Center for Public Resources. (1982). *Basic skills in the U.S. work force*. New York: Author.

Chan, L. M. W., & Mountain, D. C. (1987). Technological change and economies of scale in Canadian financial institutions: A selection from competing hypotheses. *Journal of Economics and Business, 39*, 57-66.

Child, J. (1981). Culture, contingency and capitalism in the cross-national study of organizations. *Research in Organizational Behavior, 3*, 303-356.

Chusmir, L. H., & Koberg, C. S. (1988). Religion and attitudes toward work: A new look at an old question. *Journal of Organizational Behavior, 9*, 251-262.

Clarke, F. H., & Darrough, M. N. (1983). Optimal employment contracts in a principal-agent relationship. *Journal of Economic Behavior & Organization, 4*, 69-90.

Cooper, C. L., & Robertson, I. T. (Eds.). (1986). *International review of industrial and organizational psychology*. Chichester, United Kingdom: John Wiley.

Cooper, E. A., & Barrett, G. V. (1984). Equal pay and gender: Implications of court cases for personal practices. *Academy of Management Review, 9*, 84-94.

Cooper McGovern, L. (1988). Formal end-user training declines while PC-based hardware support rises. *PC Week, 5*(24), 122.
Crosby, F. J. (1982). *Relative deprivation & working women.* New York: Oxford University Press.
Davidson, C., & Reich, M. (1988). Income inequality: An inter-industry analysis. *Industrial Relations, 27,* 263-286.
Davies, A. (1986). *Industrial relations and new technology.* London: Croom Helm.
Davis-Blake, A., & Pfeffer, J. (1989). Just a mirage: The search for dispositional effects in organizational research. *Academy of Management Review, 14,* 385-400.
DDI (Development Dimensions International). (1984, April). *American Assembly of Collegiate Schools of Business Project, Phase III, Summary Report.* Pittsburgh, PA: Author.
Deci, E. L. (1975). *Intrinsic motivation.* New York: Plenum.
de Pay, D. (1989). *Die Organisation von Innovation: Ein transaktionskostentheoretischer Ansatz* [The organization of innovation]. Wiesbaden: Gabler.
Dewar, R. D., & Dutton, J. E. (1986). The adoption of radical and incremental innovations: An empirical analysis. *Management Science, 32,* 1422-1433.
Dierkes, M., & von Thienen, V. (1984). *Kein Ende der Akzeptanzschwierigkeiten moderner Technik? Zum Zusammenhang der informationstechnischen Entwicklung und ihrer Akzeptanz* (Report No. P84-2). Berlin, West Germany: Science Center Berlin.
Dobbins, G. H. (1985). Effects of gender on leaders' responses to poor performers: An attributional interpretation. *Academy of Management Journal, 28,* 587-598.
Doswell, A. (1983). *Office automation.* New York: John Wiley.
Drazin, R., & Van de Ven, A. H. (1985). Alternative forms of fit in contingency theory. *Administrative Science Quarterly, 30,* 514-539.
Driver, M. J. (1979). Career concept and career management in organizations. In C. J. Cooper (Ed.), *Behavioral problems in organizations* (pp. 79-139). Englewood Cliffs, NJ: Prentice-Hall.
Driver, M. J. (1985). Demographic and societal factors affecting the linear career crisis. *Canadian Journal of Administrative Studies, 2,* 245-263.
Dubin, R. (Ed.). (1976). *Handbook of work, organization, and society.* Chicago: Rand McNally.
Dyer, L., & Schwab, D. P. (1982). Personnel/human resource management research. In T. A. Kochan, D. J. B. Mitchell, & L. Dyer (Eds.), *Industrial relations research in the 1970s: Review and appraisal* (pp. 187-220). Madison, WI: Industrial Relations Research Associates.
Edwards, L. N. (1988). Equal employment opportunity in Japan: A view from the West. *Industrial and Labor Relations Review, 41,* 240-250.
Ellis, P. (1984). Office planning and design: The impact of organizational change due to advanced information technology. *Behaviour and Information Technology, 3,* 221-233.
Emery, F. E., & Trist, E. L. (1960). Socio-technical systems. In *Management science models and techniques* (Vol. 2). London: Pergamon.
Emmerij, L. (1983). Paid educational leave: A proposal based on the Dutch case. In H. M. Levin & H. G. Schuetze (Eds.), *Financing recurrent education.* Beverly Hills, CA: Sage.
Eriksson, I. V. (1990). Educating end-users to make more effective use of information systems. In U. E. Gattiker (Ed.), *Technological innovation and human resources: Vol. 2. End-user training.* New York: De Gruyter.
Ettlie, J. E. (1988). *Taking charge of manufacturing: How companies are combining technological and organizational innovations to compete successfully.* San Francisco: Jossey-Bass.
Ettlie, J. E., Bridges, W. P., & O'Keefe, R. D. (1984). Organizational strategy and structural differences for radical vs. incremental innovation. *Management Science, 30,* 682-695.

References

Fay, J. A., & Medoff, J. L. (1985). Labor and output over the business cycle: Some direct evidence. *American Economic Review, 75,* 638-655.

Feldman, D. C. (1981). The multiple socialization of organization members. *Academy of Management Review, 6,* 309-318.

Fellman, T., Bräuninger, U., Gierer, R., & Grandjean, E. (1982). An ergonomic evaluation of VDTs. *Behaviour and Information Technology, 1,* 69-80.

Festinger, L. (1954). A theory of social comparison processes. *Human Relations, 7,* 117-140.

Feuer, M., Glick, H., & Desai, A. (1987). Is firm-sponsored education viable? *Journal of Economic Behavior and Organization, 8,* 121-136.

Fiorito, J., Lowman, C., & Nelson, F. D. (1987). The impact of human resource policies on union organizing. *Industrial Relations, 26,* 113-126.

Fishbein, M. (Ed.). (1967). *Readings in attitude theory and measurement.* New York: John Wiley.

Fligstein, N. (1987). The intraorganizational power struggle: Rise of finance personnel to top leadership in large corporations, 1919-1979. *American Sociological Review, 52,* 44-58.

Floyd, S. W. (1988). A micro level model of information technology use by managers. In U. E. Gattiker & L. Larwood (Eds.), *Studies in technological innovation and human resources: Managing technological development* (Vol. 1, pp. 123-142). Hawthorne, NY: De Gruyter

Form, W., & McMillen, D. B. (1983). Women, men, and machines. *Work and Occupations, 10,* 147-178.

Fossum, J. A., & Fitch, M. K. (1985). The effects of individual and contextual attributes on the sizes of recommended salary increases. *Personnel Psychology, 38,* 587-602.

Foster, L. W., & Flynn, D. M. (1984). Management information technology: Its effects on organizational forms and function. *MIS Quarterly, 8,* 229-236.

Friedman, A. (1987). Specialist labour in Japan: Computer skilled staff and the subcontracting system. *British Journal of Industrial Relations, 25,* 353-369.

Gagliardi, P. (1986). The creation and change of organizational cultures: A conceptual framework. *Organization Studies, 7,* 117-134.

Gattiker, U. E. (1984). Managing computer-based office information technology: A process model for management. In H. W. Hendrick & O. Brown, Jr. (Eds.), *Human factors in organizational design* (pp. 395-403). Amsterdam, the Netherlands: Elsevier Science.

Gattiker, U. E. (1985a). Human resources utilization and career management. In *Proceeding of the national meeting of the Association of Human Resources Management and Organizational Behavior* (pp. 458-471). New York: Maximilian.

Gattiker, U. E. (1985b). Organizational careers: Developing a construct of career success (Doctoral dissertation, Claremont Graduate School, 1985). *Dissertation Abstracts International, 45,* 4409A.

Gattiker, U. E. (1987a). *Computer application skills and the effect on the use of technology by the end-user* (Working paper). Lethbridge, Canada: University of Lethbridge.

Gattiker, U. E. (1987b). *Social comparisons at work: Wage inequality and employees' disposition about financial success* (Working paper). Lethbridge, Canada: University of Lethbridge.

Gattiker, U. E. (1988a). Technological adaptation: A typology for strategic human resource management. *Behaviour & Information Technology, 7,* 345-359.

Gattiker, U. E. (1988b). *Western Canada and the Western United States: Similarities and differences in organizational commitment* (Working paper). Lethbridge, Canada: University of Lethbridge.

Gattiker, U. E. (1988c). *Western Canada and the Western United States: Similarities and differences in perception of career success* (Working paper). Lethbridge, Canada: University of Lethbridge.

Gattiker, U. E. (1989). *Pay differentials and perceptions of career success: Comparing respondents from Western Canada and the Western U.S.* Paper presented at the 53rd annual meeting of the Academy of Management, Washington, DC.

Gattiker, U. E. (1990a). Individual differences and acquiring computer literacy: Are women more efficient than men? In U. E. Gattiker (Ed.), *Studies in technological innovation and human resources: Vol. 2. End-user training.* Hawthorne, NY: De Gruyter.

Gattiker, U. E. (1990b). *Human resources and computer technology: Current issues and emerging trends in end user training* (Working paper). Lethbridge, Canada: University of Lethbridge.

Gattiker, U. E. (1990c). *End-user skills and training: An exploratory assessment* (Working paper). Lethbridge, Canada: University of Lethbridge.

Gattiker, U. E., Gutek, B. A., & Berger, D. E. (1988). Office technology and employee attitudes. *Social Science Computer Review, 6,* 327-340.

Gattiker, U. E., & Hlavka, A. (1989). *Computer attitudes and learning performance: Issues for management education and training.* Paper presented at the 53rd annual meeting of the Academy of Management, Washington, DC.

Gattiker, U. E., & Larwood, L. (1986a). Subjective career success: A study of managers and support personnel. *Journal of Business and Psychology, 1,* 78-94.

Gattiker, U. E., & Larwood, L. (1986b). Resistance to change: Reactions to workplace computerization in offices. In B. Gold (Chair), *Computerization and productivity* (Symposium conducted at the annual meeting of the TIMS/ORSA, Los Angeles).

Gattiker, U. E., & Larwood, L. (1988). A study of potential predictors for corporate managers' objective and subjective career success. *Human Relations, 41,* 569-591.

Gattiker, U. E., & Larwood, L. (1990). Predictors for career achievement in the corporate hierarchy. *Human Relations, 43,*

Gattiker, U. E., & Nelligan, T. (1988). Computerized offices in Canada and the United States: Investigating dispositional similarities and differences. *Journal of Organizational Behaviour, 9,* 77-96.

Gibbs, B., Keen, K., & Lucas, R. (1988). Innovation and human resource productivity in Canada: A comparison of "high" and "low" technology industries. In U. E. Gattiker & L. Larwood (Eds.), *Studies in technological innovation and human resources: Managing technological development* (Vol. 1, pp. 93-120). Hawthorne, NY: De Gruyter.

Gielen, U. P. (1982). A comparison of ideal self-ratings between American and German university students. In L. L. Adler (Ed.), *Cross-cultural research at issue* (pp. 275-288). New York: Academic Press.

Glick, H. A., & Feuer, M. J. (1984). Employer-sponsored training and the governance of specific human capital investments. *Quarterly Review of Economics and Business, 24*(2), 91-103.

Goethals, G. G. (1986). Social comparison theory: Psychology from the lost and found. *Personality and Social Psychology Bulletin, 12,* 261-278.

Gold, B. (1983). Strengthening managerial approaches to improving technological capabilities. *Strategic Management Journal, 4,* 209-220.

Gold, B., Rosegger, G., & Boylan, M. (1980). *Evaluating technological innovations.* Lexington, MA: Lexington.

Goldstone, J. A. (1987). Cultural orthodoxy, risk, and innovations: The divergence of East and West in the early modern world. *Sociological Theory, 5,* 119-135.

Goodman, P. S. (1974). An examination of referents used in the evaluation of pay. *Organizational Behavior and Human Performance, 12,* 170-195.

Greenberg, J. (1982). Approaching equity and avoiding inequity in groups and organizations. In J. Greenberg & R. L. Cohen (Eds.), *Equity and justice in social behavior* (pp. 389-435). New York: Academic Press.

Griffin, R. W. (1983). Objective and social sources of information in task redesign: A field experiment. *Administrative Science Quarterly, 28*, 184-200.

Griffin, R. W., Bateman, T. S., Wayne, S. J., & Head, T. C. (1987). Objective and social factors as determinants of task perceptions and responses: An integrated perspective and empirical investigation. *Academy of Management Journal, 30*, 501-523.

Gutek, B. A. (1983). Women's work in the office of the future. In J. Zimmerman (Ed.), *The technological woman*. New York: Praeger.

Gutek, B. A. (1986). *Sex and the workplace*. San Francisco: Jossey-Bass.

Gutek, B. A., & Bikson, T. K. (1985). Differential experiences of men and women in computerized offices. *Sex Roles, 13*, 123-136.

Gutek, B. A., & Larwood, L. (1987). Information technology and working women in the USA. In M. J. Davidson & C. L. Cooper (Eds.), *Women and technology*. Chichester, United Kingdom: John Wiley.

Gutek, B. A., Stromberg, A. H., & Larwood, L. (Eds.). (1988). *Women and work: An annual review* (Vol. 3). Newbury Park, CA: Sage.

Hackman, J. R., & Oldham, G. R. (1980). *Work redesign*. Reading, MA: Addison-Wesley.

Hage, J. (1972). *Techniques and problems of theory construction in sociology*. New York: John Wiley.

Halaby, C. N. (1979). Sexual inequality in the workplace: An employer-specific analysis of pay differences. *Social Science Research, 8*, 79-104.

Hall, R. E. (1982). The importance of lifetime jobs in the U.S. economy. *American Economic Review, 72*, 716-724.

Hall, R. H. (1986). *Dimensions of work*. Beverly Hills, CA: Sage.

Hambrick, D. C. (1981). Environment, strategy and power within top management teams. *Administrative Science Quarterly, 26*, 253-276.

Harris and Associates. (1980). *The Steelcase national study of office environments no. II: Comfort and productivity in the office of the 80's*. Grand Rapids, MI: Steelcase.

Hedge, A. (1982). The open-plan office: A systematic investigation of employee reactions to their work environment. *Environment and Behavior, 14*, 519-542.

Hefti, R. (1983). Die Zukunft begann 1972 . . . [The future started in 1972 . . .]. *SKA Bulletin*, (10), 47-48.

Heneman, H. G., III, & Schwab, D. P. (1979). Work and rewards theory. In D. Yoder & H. G. Heneman, Jr. (Eds.), *ASPA handbook of personnel and industrial relations* (pp. 6.1-6.22). Washington, DC: Bureau of National Affairs.

Hickson, D. J. (1988). Offence and defence: A symposium with Hinings, Clegg, Child, Aldrich, Karpik, and Donaldson. *Organization Studies, 9*, 1.

Hickson, D. J., Hinings, C. R., McMillan, C. J., & Schwitter, J. P. (1974). The culture-free context of organization structure: A trinational comparison. *Sociology, 8*, 59-80.

Hickson, D. J., & McMillan, C. J. (Eds.). (1981). *Organizations and nation: The Aston program IV*. Westmead, United Kingdom: Gower.

Hickson, D. J., McMillan, C. J., Azumi, K., & Horvath, D. (1979). Grounds for comparative organization theory: Quicksand or hard core? In C. J. Lammers & D. J. Hickson (Eds.), *Organizations alike and unlike* (pp. 25-41). London: Routledge & Kegan Paul.

Hinrichs, J. R. (1976). Personnel training. In M. D. Dunnette (Ed.), *Handbook of industrial and organizational psychology* (pp. 829-860). Chicago: Rand McNally.

Hirschheim, R. A. (1985). *Office automation: A social and organizational perspective.* Chichester, United Kingdom: John Wiley.

Hitt, M. A., Ireland, R. D., & Goryunov, I. Y. (1988). The context of innovation: Investment in R & D and firm performance. In U. E. Gattiker & L. Larwood (Eds.), *Studies in technological innovation and human resources: Managing technological development* (Vol. 1, pp. 73-92). Hawthorne, NY: De Gruyter.

Hodson, R. (1983). *Workers' earnings and corporate economic structure.* New York: Academic Press.

Hofstede, G. (1980). *Culture's consequences: International differences in work-related values.* Beverly Hills, CA: Sage.

Hofstede, G. (1984). *Culture's consequences* (abridged ed.). Beverly Hills, CA: Sage.

Hogan, R., Carpenter, B. N., Briggs, S. R., & Hansson, R. O. (1985). Personality assessment and personnel selection. In H. J. Bernardin & D. A. Bownas (Eds.), *Personality assessment in organizations* (pp. 21-52). New York: Praeger.

Hogarth, R. M. (1980). *Judgement and choice: The psychology of decision.* New York: John Wiley.

Hollenbeck, J. R., Ilgen, D. R., Ostroff, C., & Vancouver, J. B. (1987). Sex differences in occupational choice, pay and worth: A supply-side approach to understanding the male-female wage gap. *Personnel Psychology, 40*, 715-743.

Homans, G. C. (1980). Discovery and the discovered in social theory. In H. M. Blalock (Ed.), *Sociological theory and research* (pp. 17-22). New York: Free Press.

Hoskisson, R. E., & Galbraith, C. S. (1985). The effect of quantum vs. incremental M-form reorganization on performance: A time-series exploration of intervention dynamics. *Journal of Management, 11*(3), 55-70.

Howard, A. (1986). College experiences and managerial performance. *Journal of Applied Psychology, 71*, 530-552.

Hrebiniak, L. G., & Alutto, J. (1972). Personal and role-related factors in the development of organizational commitment. *Administrative Science Quarterly, 17*, 555-572.

Hrebiniak, L. G., & Joyce, W. F. (1984). *Implementing strategy.* New York: Macmillan.

Hrebiniak, L. G., & Joyce, W. F. (1985). Organizational adaptation: Strategic choice and environmental determinism. *Administrative Science Quarterly, 30*, 336-349.

Ichniowski, C. (1986). The effects of grievance activity on productivity. *Industrial and Labor Relations Review, 40*, 75-89.

IG-Metall. (1984). *Action program: Work and technology. "People must stay!"* Frankfurt: Industriegewerkschaft Metall fuer die Bundesrepublik Deutschland.

Ilgen, D. R., & Feldman, J. M. (1983). Performance appraisal: A process focus. In L. L. Cummings & B. Staw (Eds.), *Research in organizational behavior* (pp. 141-197). Greenwich, CT: JAI.

Jacklin, C. N. (1989). Female and male: Issues of gender. *American Psychologist, 44*, 127-133.

Jackofsky, E. F., & Peters, L. H. (1987). Part-time versus full-time employment status differences: A replication and extension. *Journal of Occupational Behaviour, 8*, 1-9.

Jacoby, S. M., & Mitchell, D. J. B. (1986). Management attitudes toward two-tier pay plans. *Journal of Labour Research, 7*, 221-237.

Jain, H. C. (1985). Industrial-relations practices of Japanese multinationals and their subsidiaries in Canada: Are they transferable to Canadian firms? *Canadian Journal of Administrative Sciences, 2*, 77-94.

James, L. R., & Tetrick, L. E. (1986). Confirmatory analytic tests of three causal models relating job perceptions to job satisfaction. *Journal of Applied Psychology, 71*, 77-82.

References

Jones, J. W., & Lavelli, M. A. (1986). Essential computing skills needed by psychology students seeking careers in business. *Journal of Business and Psychology, 1,* 163-167.

Kahn, R. L. (1981). Work and health: Some psychosocial effects of advanced technology. In B. Gardell & G. Johansson (Eds.), *Working life* (pp. 17-37). Chichester, United Kingdom: John Wiley.

Kalleberg, A. L., & Berg, I. (1987). *Work and industry.* New York: Plenum.

Kamm, J. B. (1987). *An integrative approach to managing innovation.* Lexington, MA: Lexington.

Katz, D., & Kahn, R. L. (1978). *The social psychology of organizations* (2nd ed.). New York: John Wiley.

Katz, D., Kahn, R. L., & Adams, S. (Eds.). (1980). *The study of organizations* (pp. 537-554). San Francisco: Jossey-Bass.

Kaufman, H. G. (1989). Obsolescence of technical professionals: A measure and a model. *Applied Psychology: An International Review, 38,* 73-85.

Kauppinen-Toropainen, K., Kandolin, J., & Haavio-Mannila, E. (1988). Sex segregation of work in Finland and the quality of women's work. *Journal of Organizational Behavior, 9,* 15-27.

Kemp, A. A., & Beck, E. M. (1986). Equal work, unequal pay. *Work and Occupations, 13,* 324-347.

Kerr, I L. (1985). Diversification strategies and managerial rewards: An empirical study. *Academy of Management Journal, 28,* 155-179.

Kerr, S. (1975). On the folly of rewarding A while hoping for B. *Academy of Management Journal, 18,* 769-783.

Kimberly, J. R. (1981). Managerial innovation. In P. C. Nystrom & W. H. Starbuck (Eds.), *Handbook of organizational design* (Vol. 1, pp. 84-104). New York: Oxford University Press.

Kimberly, J. R. (1987). Organizational and contextual influences on the diffusion of technological innovation. In J. M. Pennings & A. Buitendam (Eds.), *New technology as organizational innovation* (pp. 237-259). Cambridge, MA: Ballinger.

Kingstrom, P. O., & Mainstone, L. E. (1985). An investigation of the rater-ratee acquaintance and rater bias. *Academy of Management, 28,* 641-653.

Kirmeyer, S. L., & Shirom, A. (1986). Perceived job autonomy in the manufacturing sector: Effects of unions, gender, and substantive complexity. *Academy of Management Journal, 29,* 832-840.

Klein, K. J., Hall, R. J., & LaLiberte, M. (1990). Training and the organizational consequences of technological change: A case study of computer-aided design and drafting. In U. E. Gattiker (Ed.), *Technological innovation and human resources* (Vol. 2, pp. 7-36). Berlin and New York: Walter de Gruyter.

Kleingartner, A., & Anderson, C. S. (Eds.). (1987). *Human resource management in high technology firms.* Lexington, MA: Lexington.

Kling, R., & Iacono, S. (1984). Computing as an occasion for social control. *Journal of Social Issues, 40,* 77-96.

Knipper, H-J. (1989). Entlassungen nur bei hohen Abfindungen möglich [Lay-offs only with high severance packages possible]. *Handelsblatt, 143,* 5.

Kohn, M. L. (1989). Cross-national research as an analytic strategy. In M. L. Kohn (Ed.), *Cross-national research in sociology* (pp. 77-102). Newbury Park, CA: Sage.

Kohn, M. L., & Schooler, C., in collaboration with Miller, J., Miller, K. A., Schoenbach, C., & Schoenberg, R. (1983). *Work and personality: An inquiry into the impact of social stratification.* Norwood, NJ: Ablex.

Kotter, J. P. (1982). *The general managers.* New York: Free Press.

Kraft, P., & Dubnoff, S. (1986). Job content, fragmentation, and control in computer software work. *Industrial Relations, 25*, 184-196.

Kram, K. E. (1985). *Mentoring at work: Developmental relationships in organizational life.* Glenview, IL: Scott, Foresman.

Kram, K. E., & Isabella, L. A. (1985). Mentoring alternatives: The role of peer relationships in career development. *Academy of Management Journal, 28*, 110-132.

Krebsbach-Gnath, C., Ballerstedt, E., Frenzel, U., Bielenski, H., Büchteman, C. F., & Bengelmann, D. (1983). *Frauenbeschäftigung und neue Technologien* [Employment of women and new technologies]. Munich: R. Oldenbourg.

Kuechler, M. (1987). The utility of surveys for cross-national research. *Social Science Research, 16*, 229-244.

Kugler, P. (1988). *Lohndiskriminierung in der Schweiz* [Wage discrimination in Switzerland]. Berne: Eidgenoessisches Justiz – und Polizeidepartement – Arbeitsgruppe "Lohngleichheit," Forschungsbericht Nr. 1.

Kuhn, T. S. (1970). *The structure of scientific revolutions* (2nd, enlarged ed.). Chicago: University of Chicago Press.

Landau, J., & Hammer, T. H. (1986). Clerical employees' perceptions of intraorganizational career opportunities. *Academy of Management Journal, 29*, 385-404.

Landy, F. J. (1985). *Psychology and work behavior* (3rd ed.). Homewood, IL: Dorsey.

Larwood, L., & Gattiker, U. E. (1986). A comparison of the career paths used by successful women and men. In B. A. Gutek & L. L. Larwood (Eds.), *Women's career development* (pp. 129-156). Beverly Hills, CA: Sage.

Larwood, L., Gutek, B. A., & Gattiker, U. E. (1984). Perspectives on institutional discrimination and resistance to change. *Group & Organization Studies, 9*, 333-352.

Larwood, L., Stromberg, A. H., & Gutek, B. A. (Eds.). (1985). *Women and work: An annual review* (Vol. 1). Beverly Hills, CA: Sage.

Lawler, E. E. (1981). *Pay and organization development.* Reading, MA: Addison-Wesley.

Lazarus, R. S. (1982). Thoughts on the relations between ambition and cognition. *American Psychologist, 37*, 1019-1024.

Lazarus, R. S. (1984). On the primacy of cognition. *American Psychologist, 39*, 124-129.

Leibowitz, Z. B., Farren, C., & Kaye, B. L. (1986). *Designing career development systems.* San Francisco: Jossey-Bass.

Lengnick-Hall, C. A. (1986). Technology advances in batch production and improved competitive position. *Journal of Management, 12*, 75-90.

Leontief, W., & Duchin, F. (1986). *The future impact of automation on workers.* New York: Oxford University Press.

Lerne und arbeite [Learn and work]. (1989, May 19). *Wirtschaftswoche,* pp. 98-113.

Levin, M. H., & Schuetze, H. G. (Eds.). (1983). *Financing recurrent education.* Beverly Hills, CA: Sage.

Lieberman, M. A., Selig, G. J., & Walsh, J. J. (1982). *Office automation.* Beverly Hills, CA: Sage.

Link, N. A., & Tassey, G. (1987). *Strategies for technology-based competition.* Lexington, MA: Lexington.

Lippitt, G. L., Langseth, P., & Mossop, J. (1985). *Implementing organizational change.* San Francisco: Jossey-Bass.

Locke, E. A. (1968). Toward a theory of task motivation and incentives. *Organizational Behavior and Human Performance, 3*, 157-189.

References

Locke, E. A. (1976). The nature and causes of job satisfaction. In M. D. Dunnette (Ed.), *Handbook of industrial and organizational psychology*. Chicago: Rand McNally.

Long, R. J. (1987). *New office information technology: Human and managerial implications.* London: Croom Helm.

Lorence, J. (1987). Intraoccupational earnings inequality: Human capital and institutional determinants. *Work and Occupations, 14*, 236-260.

Loscocco, K. A. (1989). The instrumentally oriented factory worker: Myth or reality? *Work and Occupations, 16*, 3-25.

Louis, M. R. (1980). Surprise and sense making: What newcomers experience in entering unfamiliar organizational settings. *Administrative Science Quarterly, 25*, 226-251.

Lowe, G. S., & Krahn, H. (1988). *Computer skills end use among high school and university graduates* (Working paper). Alberta, Canada: University of Alberta.

Luthans, F., McCaul, H. S., & Dodd, N. G. (1985). Organizational commitment: Analysis of antecedents. *Academy of Management Journal, 28*, 213-219.

Mankin, D., Bikson, T. K., & Gutek, B. A. (1984). Factors in successful implementation of computer-based office information systems: A review of the literature with suggestions for OBM research. *Journal of Organizational Behavior, 6*, 1-20.

Mansfield, R. (1973). Bureaucracy and centralization: An examination of organizational structure. *Administrative Science Quarterly, 18*, 77-88.

March, J. G., & Simon, H. A. (1958). *Organizations.* New York: John Wiley.

Mason, N. A., & Belt, J. A. (1986). Effectiveness of specificity in recruitment advertising. *Journal of Management, 12*, 425-432.

Maurice, M., Sorge, A., & Warner, M. (1980). Societal differences in organizing manufacturing units: A comparison of France, West Germany, and Great Britain. *Organization Studies, 1*, 59-86.

McClelland, D. C. (1975). *Power: The inner experience.* New York: John Wiley.

Meyer, J. W., & Scott, W. R. (1983). *Organizational environment: Ritual and rationality.* Beverly Hills, CA: Sage.

Meyer, M. W., Stevenson, W., & Webster, S. (1985). *Limits to bureaucratic growth.* New York: De Gruyter.

Miles, R., & Snow, C. C. (1978). *Organizational strategy structure and process.* New York: McGraw-Hill.

Miller, D., & Droege, C. (1986). Psychological and traditional determinants of structure. *Administrative Science Quarterly, 31*, 539-560.

Miller, D., & Friesen, P. H. (1980). Momentum and revolution in organizational adaptation. *Academy of Management Journal, 23*, 591-614.

Miller, D., & Friesen, P. H. (1986). Porter's (1980) generic strategies and performance: An empirical examination with American data. Part I: Testing Porter. *Organization Studies, 7*, 37-55.

Miller, G. A. (1987). Meta-analysis and the culture-free hypothesis. *Organization Studies, 8*, 309-325.

Miller, H. D., & Terborg, J. T. (1979). Job attitudes of part-time and full-time employees. *Journal of Applied Psychology, 64*, 380-386.

Mincer, J. (1974). *Schooling, experience, and earnings.* New York: National Bureau of Economic Research.

Minor, F. J. (1986a, August). *Assessment of a computer-based career planning information system.* Paper presented at the 50th annual meeting of the Academy of Management, Chicago.

Minor, F. J. (1986b). Computer applications in career development. In D. T. Hall (Ed.), *Career development in organizations* (pp. 202-235). San Francisco: Jossey-Bass.

Mitroff, I. I. (1983). *Stakeholders of the organizational mind*. San Francisco: Jossey-Bass.

Mobley, W. H. (1982). *Employee turnover: Causes, consequences, and control*. Reading, MA: Addison-Wesley.

Montgomery, M. (1988). Hours of part-time and full-time workers at the same firm. *Industrial Relations, 27*, 394-406.

Morgall, J. (1983). Typing our way to freedom: Is it true that office technology can liberate women? *Behaviour and Information Technology, 2*, 215-226.

Morgan, G. (1986). *Images of organization*. Beverly Hills, CA: Sage.

Möschel, W. (1989). Postreform im Zwielicht [Postal reform in the twilight zone]. *WiSt, 18*(4), 173-179.

Mount, M. K., & Thompson, D. E. (1987). Cognitive categorization and quality of performance ratings. *Journal of Applied Psychology, 72*, 240-246.

Mowday, R. T., Porter, L. W., & Steers, R. M. (1982). *Employee organization linkages*. New York: Academic Press.

Murphy, K. R., Gannett, B. A., Herr, B. M., & Chen, J. A. (1986). Effects of subsequent performance on evaluations of previous performance. *Journal of Applied Psychology, 71*, 427-431.

Neal, A. C. (1987). Co-determination in the Federal Republic of Germany: An external perspective from the United Kingdom. *British Journal of Labour Relations, 25*, 227-245.

Newton, R. D. (1984). Job satisfaction and somatic complaints among computer aided design drafters (Doctoral dissertation, Claremont Graduate School, 1984). *Dissertation Abstracts International, 45*(4), 1311B-1312B.

Nolan, Norton & Co. (1987). *Managing personal computers in the large organization*. Lexington, MA: Author.

Nord, W. R., & Tucker, S. (1987). *Implementing routine and radical innovations*. Boston: D. C. Heath.

Nuti, D. M. (1987). Profit-sharing and employment: Claims and overclaims. *Industrial Relations, 26*, 18-29.

Nutt, P. C. (1986). Evaluating MIS design principles. *MIS Quarterly, 10*, 139-156.

NZZ. (1988, January 9). Probleme der flexiblen Arbeit in den Niederlanden [Problems with flexible work in Holland]. *Neue Zürcher Zeitung*, Fernausgabe Nr. 5, p. 15.

NZZ. (1989a, April 8). 37'000 neue Arbeitsplätze in der Wirtschaft [37,000 new jobs in the economy]. *Neue Zürcher Zeitung*, Fernausgabe Nr. 80, p. 13.

NZZ. (1989b, April 8). Deutsche Metallarbeiter rüsten zum Grosskonflikt [German metal workers prepare for major conflict]. *Neue Zürcher Zeitung*, Fernausgabe Nr. 80, p. 13.

NZZ. (1989c, December 14). Massive Forderungender IG Metall [Massive demands by IG Metall]. *Neue Zürcher Zeitung*, Fernausgabe Nr. 290, p. 19.

OECD (Organization for Economic Cooperation and Development). (1973). *Recurrent education: A strategy for lifelong learning*. Paris: Author.

OECD (Organization for Economic Cooperation and Development). (1982). *Meeting of management experts on the adjustment of working time: Economic and employment implications*. Paris: Author.

Okubayashi, K. (1986). Recent problems of Japanese personnel management. *Labour and Society, 11*(1), 17-37.

Okubayashi, K. (1987). Work content and organisational structure of Japanese enterprises under microelectronic innovation. *Annals of the School of Business Administration* (Kobe University), *31*, 32-52.

Oldham, G. R., Kulik, C. T., Ambrose, M. L., Stepina, L. P., & Brand, J. F. (1986). Relations between job facet comparisons and employee reactions. *Organizational Behavior and Human Decision Processes, 38*, 28-47.

Oldham, G. R., Kulik, C. T., Stepina, L. P., & Ambrose, M. L. (1986). Relations between situational factors and the comparative referents used by employees. *Academy of Management Journal, 29*, 599-608.

Oldham, G. R., Nottenburg, G., Wright Kassner, M., Ferris, G., Fedor, D., & Masters, M. (1982). The selection and consequences of job comparisons. *Organizational Behavior and Human Performance, 29*, 84-111.

Oldham, G. R., & Rotchford, N. L. (1983). Relationships between office characteristics and employee reactions: A study of the physical environment. *Administrative Science Quarterly, 28*, 542-556.

Olian, J. D., & Rynes, S. L. (1984). Organizational staffing: Integrating practice with strategy. *Industrial Relations, 23*, 170-183.

O'Reilly, A. (1983). The use of information in organizational decision making: A model and some propositions. *Research in Organizational Behavior, 5*, 103-139.

O'Reilly, C. A., & Caldwell, D. F. (1981). The commitment and job tenure of new employees: Some evidence of postdecisional justification. *Administrative Science Quarterly, 26*, 597-616.

Oskamp, S. (1977). *Attitudes and opinions*. Englewood Cliffs, NJ: Prentice-Hall.

Osterloh, M. (1986). Zum Problem der rechtzeitigen Information von Arbeitnehmervertretern in Betriebs – und Aufsichtsräten. *Arbeit und Recht, 34*, 332-340.

Osterman, P. (1984). White-collar internal labor markets. In P. Osterman (Ed.), *Internal labor markets* (pp. 163-189). Cambridge: MIT Press.

Osterman, P. (1986). Choice of employment systems in internal labor markets. *Industrial Relations, 26*, 46-67.

Panko, R. R. (1984). Office work. *Office: Technology and People, 2*, 205-238.

Panko, R. R. (1987). Directions and issues in end user computing. *INFOR – Information Systems and Operational Research, 25*, 181-197.

Parcel, T. L., & Mueller, C. W. (1983). *Ascription and labor markets: Race and sex differences in earnings*. New York: Academic Press.

Pava, C. (1983). *Managing new office technology*. New York: Free Press.

Payne, R. L., & Jones, J. G. (1987a). Social class and re-employment: Changes in health and perceived financial circumstances. *Journal of Occupational Behaviour, 8*, 175-184.

Payne, R. L., & Jones, J. G. (1987b). The effects of long-term unemployment on attitudes to employment. *Journal of Occupational Behaviour, 8*, 351-358.

Pearce, J. L., Stevenson, W. B., & Perry, J. L. (1985). Managerial compensation based on organization performance: A time series analysis of the effects of merit pay. *Academy of Management Journal, 28*, 261-278.

Pennings, J. M. (1987). Structural contingency theory: A multivariate test. *Organization Studies, 8*, 223-240.

Pennings, J. M., & Buitendam, A. (1987) (Eds.). *New technology as organizational innovation*. Cambridge, MA: Ballinger.

Penrose, J., & Seiford, L. M. (1988). Computer "personalities": A new approach to understanding user compatibility. *Journal of Business and Psychology, 3*, 74-87.

Perry, L. T., & Sandholtz, K. W. (1988). A "Liberating Form" for radical product innovation. In U. E. Gattiker & L. Larwood (Eds.), *Studies in technological innovation and human resources: Managing technological development* (Vol. 1, pp. 9-31). Hawthorne, NY: De Gruyter.

Petermann, T., & von Thienen, V. (1988). Technikakzeptanz: Zum Karriereverlauf eines Begriffes. In R. G. von Westphalen (Ed.), *Technikfolgenabschätzung als politische Aufgabe* (pp. 85-148). Opladen, FRG: Westdeutscher Verlag.

Peters, L. H., Jackofsky, E. F., & Salter, J. R. (1981). Predicting turnover: A comparison of part-time and full-time employees. *Journal of Occupational Behaviour, 2*, 89-98.

Pettigrew, A. (1985). *The awakening giant.* Oxford: Basil Blackwell.

Pfeffer, J. (1981a). Four laws of organizational research. In A. H. Van de Ven & W. F. Joyce (Eds.), *Perspectives on organization design and behaviour* (pp. 409-418). New York: John Wiley.

Pfeffer, J. (1981b). Management as symbolic action: The creation and maintenance of organizational paradigms. In L. L. Cummings & B. M. Staw (Eds.), *Research in organizational behavior* (Vol. 3, pp. 1-52). Greenwich, CT: JAI.

Pfeffer, J. (1982). *Organizations and organization theory.* Boston: Pitman.

Pfeffer, J. (1983). Organizational demography. In B. Staw & L. L. Cummings (Eds.), *Research in organizational behavior* (Vol. 5, pp. 299-357). Greenwich, CT: JAI.

Pfeffer, J., & Cohen, Y. (1984). Determinants of internal labour markets in organizations. *Administrative Science Quarterly, 29*, 550-572.

Pfeffer, J., & Davis-Blake, A. (1987). Understanding organizational wage structures: A resource dependence approach. *Academy of Management Journal, 30*, 437-455.

Pfeffer, J., & Lawler, J. (1980). Effects of job alternatives, extrinsic rewards, and behavioral commitment on attitude toward the organization. *Administrative Science Quarterly, 25*, 38-56.

Pfeffer, J., & Salancik, G. R. (1978). *The external control of organizations: A resource dependence perspective.* New York: Harper & Row.

Pfeffer, J., Salancik, G. R., & Leblebici, H. (1976). The effect of uncertainty on the use of social influence in organizational decision making. *Administrative Science Quarterly, 21*, 227-245.

Pfister, D. (1988). Technikorientiertes Informations- und Marketingmanagement von kleinen und mittleren Unternehmen [Technology focused information and marketing management of small- and medium-sized firms]. *Volkswirtschaft, 10*, 23-25.

Piore, M. J. (1986). Perspectives on labor market flexibility. *Industrial Relations, 25*, 146-166.

Piore, M. J., & Sabel, C. F. (1984). *The second industrial divide: Possibilities for prosperity.* New York: Basic Books.

Porter, L. W., Steers, R. M., Mowday, R. T., & Boulian, P. V. (1974). Organizational commitment, job satisfaction, and turnover among psychiatric technicians. *Journal of Applied Psychology, 59*, 603-609.

Porter, M. E. (1980). *Competitive strategy.* New York: Free Press.

Quan, N. T. (1984). Unionism and the size of distribution earnings. *Industrial Relations, 23*, 270-277.

Rafaeli, A., & Klimoski, R. J. (1983). Predicting sales success through handwriting analysis: An evaluation of the effects of training and handwriting sample content. *Journal of Applied Psychology, 68*, 212-217.

Ranson, S., Hinings, B., & Greenwood, R. (1980). The structuring of organizational structures. *Administrative Science Quarterly, 25*, 1-17.

Ray, C. A. (1989). Skill reconsidered. *Work and Occupations, 16*, 65-79.

Reilly, R. R., Zedeck, S., & Tenopyr, M. L. (1979). Validity and fairness of physical ability tests for predicting performance in craft jobs. *Journal of Applied Psychology, 64*, 262-274.

References

Reith, J. (1976). Group methods: Conferences, meetings, workshops, seminars. In R. L. Craig (Ed.), *Training and development handbook* (2nd ed., chap. 34). New York: McGraw-Hill.

Rodrigues, S. (1988). *New technologies and work deskilling in the Brazilian services: Some implications for job design* (Working paper). Minas Gerais, Brazil: Universidade Federal de Minas Gerais.

Rogers, T. F., & Friedman, N. S. (1980). *Printers face automation.* Lexington, MA: Lexington.

Rohmert, W., & Landau, K. (1983). *A new technique for job analysis.* New York: Taylor & Francis.

Rokeach, M. (1980). *Beliefs, attitudes, and values* San Francisco: Jossey-Bass.

Rollier, M. (1986). Changes of industrial relations at Fiat. In O. Jacobi, B. Jessop, H. Kastendiek, & M. Regini (Eds.), *Technological change, rationalisation and industrial relations* (pp. 116-133). New York: St. Martin's.

Rosegger, G. (1986). *The economics of production and innovation* (2nd ed.). Oxford: Pergamon.

Rosenbaum, J. E. (1984). *Career mobility in a corporate hierarchy.* New York: Academic Press.

Rosenbaum, J. E. (1985). Persistence and change in pay inequalities: Implications for job evaluation and comparable worth. In L. Larwood, B. A. Gutek, & A. Stromberg (Eds.), *Women and work: An annual review* (Vol. 1, pp. 115-140). Newbury Park, CA: Sage.

Roskies, E., Liker, J. K., & Roitman, D. B. (1988). Winners and losers: Employee perceptions of their company's technological transformation. *Journal of Organizational Behavior, 9,* 123-137.

Rowe, A. J., Mason, R. O., & Dickel, K. E. (1982). *Strategic management and business policy: A methodological approach.* Reading, MA: Addison-Wesley.

Roznowski, M., & Hulin, C. L. (1985). Influence of functional specialty and job technology on employees' perceptual and affective responses to their jobs. *Organizational Behavior and Human Decision Processes, 36,* 186-208.

Rubenson, K. (1983). The Swedish model. In H. M. Levin & H. G. Schuetze (Eds.), *Financing recurrent education* (pp. 237-255). Beverly Hills, CA: Sage.

Ruch, L., & Troy, N. (1986). *Textverarbeitung im Sekretariat* [Word processing in the secretariat]. Zurich: Verlag der Fachvereine (VdF).

Salancik, G. R., & Pfeffer, J. (1977). A social information processing approach to job attitudes and task design. *Administrative Science Quarterly, 23,* 224-253.

Salancik, G. R., & Pfeffer, J. (1978). *The external control of organizations: A resource dependence perspective.* New York: Harper & Row.

Scarbrough, H. (1986). The politics of technological change at British Leyland. In O. Jacobi, B. Jessop, H. Kastendiek, & M. Regini (Eds.), *Technological change, rationalisation and industrial relations* (pp. 95-115). New York: St. Martin's.

Schein, E. H. (1978). *Career dynamics: Matching individual and organizational needs.* Reading, MA: Addison-Wesley.

Schein, E. H. (1985). *Organizational culture and leadership.* San Francisco: Jossey-Bass.

Scholl, R. W., Cooper, E. A., & McKenna, J. F. (1985). Referent selection in determining equity perceptions: Differential effects on behavioral and attitudinal outcomes. *Personnel Psychology, 40,* 113-124.

Schoonhoven, C. B. (1981). Problems with contingency theory: Testing assumptions hidden within the language of contingency "theory." *Administrative Science Quarterly, 26*(3), 349-377.

Schreyögg, G. (1980). Contingency and choice in organization theory. *Organization Studies, 1,* 73-78.

Schreyögg, G. (1982). Some comments about comments: A reply to Donaldson. *Organization Studies, 3*, 73-80.

Schweizerischer Gewerkschaftsbund. (1989). *Berufsbildung dauert ein Leben lang* [Professional/vocational training lasts a lifetime]. Bern, Switzerland: The Author.

Scott, W. R. (1981). *Organizations: Rational, natural, and open systems*. Englewood Cliffs, NJ: Prentice-Hall.

Sekaran, U. (1986). Mapping bank employee perceptions of organizational stimuli in two countries. *Journal of Management, 12*, 19-30.

Selznick, P. (1957). *Leadership in administration*. New York: Harper & Row.

Sethia, N. K., & Von Glinow, M. A. (1985). Arriving at four cultures by managing the reward system. In R. H. Kilman, M. J. Saxton, R. Serpa, & associates (Eds.), *Gaining control of the corporate culture* (pp. 400-433). San Francisco: Jossey-Bass.

Shaiken, H., Herzenberg, S., & Kuhn, S. (1986). The work process under more flexible production. *Industrial Relations, 25*, 167-183.

Shapira, Z. (1987). Preference for job attributes: Tradeoffs from present position. *Industrial Relations, 26*, 146-157.

Shepelak, N. J. (1987). The role of self-explanations and self-evaluations in legitimating inequality. *American Sociological Review, 52*, 495-503.

Sheridan, J. E. (1985). A catastrophe model of employee withdrawal leading to low job performance, high absenteeism, and job turnover during the first year of employment. *Academy of Management Journal, 28*, 88-109.

Sherman, J. D., & Smith, H. L. (1984). The influence of organizational structure on intrinsic versus extrinsic motivation. *Academy of Management Journal, 27*, 877-885.

Simon, H. A. (1957). *Models of man*. New York: John Wiley.

Simon, H. A. (1979). Rational decision making in business organizations. *American Economic Review, 69*, 493-513.

Simon, H. A. (1984). On the behavioral and rational foundations of economic dynamics. *Journal of Economic Behaviour and Organization, 5*, 35-55.

Snow, C. C., & Hrebiniak, L. C. (1980). Strategy, distinctive competence and organizational performance. *Administrative Science Quarterly, 25*, 317-336.

Sorge, A. (1985). Culture's consequences. In P. Lawrence & K. Elliott (Eds.), *Introducing management* (pp. 234-244). Harmondsworth, Middlesex, United Kingdom: Penguin.

Sorge, A. (1989). An essay on technical change: Its dimensions and social and strategic context. *Organization Studies, 10*(1), 23-44.

Sorge, A., & Warner, M. (1986). *Comparative factory organisation: An Anglo-German comparison of management and manpower in manufacturing*. Aldershot, Great Britain: Gower.

Spenner, K. I. (1983). Deciphering Prometheus: Temporal change in the skill level of work. *American Sociological Review, 48*, 824-837.

Spulber, D. F. (1985). Risk sharing and inventories. *Journal of Economic Behavior and Organization, 6*, 55-68.

Stark, D. (1986). Rethinking internal labour markets: New insights from a comparative perspective. *American Sociological Review, 51*, 492-504.

Staw, B. M., Bell, N. E., & Clausen J. A. (1986). The dispositional approach to job attitudes: A lifetime longitudinal test. *Administrative Science Quarterly, 31*, 56-77.

Staw, B. M., & Ross, J. (1985). Stability in the midst of change: A dispositional approach to job attitudes. *Journal of Applied Psychology, 70*, 469-480.

Stellman, J. M., Klitzman, S., Gordon, G. C., & Snow, B. R. (1987). Work environment and the well-being of clerical and VDT workers. *Journal of Occupational Behaviour, 8*, 95-114.

References

Stevenson, H. H. (1976). Defining corporate strengths and weaknesses. *Sloan Management Review, 17*, 51-68.

Stinchcombe, A. L. (1963). Some empirical consequences of the Davis-Moore theory of stratification. *American Sociological Review, 38*, 805-808.

Stromberg, A. H., Larwood, L., & Gutek, B. A. (Eds.). (1987). *Women and work: An annual review* (Vol. 2). Newbury Park, CA: Sage.

Swaroff, P. G., Barclay, L. A., & Bass, A. R. (1985). Recruiting sources: Another look. *Journal of Applied Psychology, 70*, 720-728.

Swidler, A. (1986). Culture in action: Symbols and strategies. *American Sociological Review, 51*, 273-286.

Sydow, J. (1985). *Organisationsspielraum und Büroautomation* [Organizational flexibility and office automation]. New York: De Gruyter.

Sydow, J. (1987). Office automation: An organizational perspective. In M. Frese, E. Ulich, & W. Dzida (Eds.), *Psychological issues of human computer interaction in the workplace* (pp. 59-80). Amsterdam: North-Holland.

Tachibanaki, T. (1987). The determination of the promotion process in organizations and of earnings differentials. *Journal of Economic Behavior and Organization, 8*, 603-616.

Tanaka, H. (1988a). On the geographic study of pilgrimage places. In S. M. Bhardwaj & G. Rinschede (Eds.), *Geographia Religionum: Interdisziplinare Schriftenreihe zur Religionsgeographie: Vol. 4. Pilgrimage in world religions*. Berlin: Dietrich Reimer.

Tanaka, H. (1988b). *Personality in industry: The human side of a Japanese enterprise*. London: Frances Pinder.

Taylor, R. P. (Ed.). (1980). *The computer in the school: Tutor, tool, tutee*. New York: Teachers College Press.

TCO (The Central Organisation of Salaried Employees in Sweden). (1986). *Screen checker*. Stockholm: Author.

Thorndike, E. L. (1913). *Educational psychology* (Vols. 1, 2). New York: Columbia University, Teachers College.

Tiedeman, D. V., & O'Hara, R. P. (1963). *Career development: Choice and adjustment*. Princeton, NJ: College Entrance Examination Board.

Tornatzky, L. G. (1986). Technological change and the structure of work. In M. S. Pallak & R. O. Perloff (Eds.), *Psychology and work: Productivity, change, and employment* (pp. 55-83). Washington, DC: American Psychological Association.

Treacy, M. E., & Index Group, Inc. (1989). *The costs of network ownership*. Cambridge, MA: Index Group, Inc.

Triandis, H. C. (1977). Cross-cultural social and personality psychology. *Personality and Social Psychology Bulletin, 56*, 143-158.

Tubbs, M. E. (1986). Goal setting: A meta-analytic examination of the empirical evidence. *Journal of Applied Psychology, 71*, 474-483.

Turner, J. A. (1980). *Computers in bank clerical functions: Implications for productivity and the quality of working life*. Ann Arbor, MI: University Microfilms International.

UBS (Union Bank of Switzerland). (1985, June). Labor costs and productivity by international comparison. *UBS Business Facts and Figures*, pp. 6-7.

UBS (Union Bank of Switzerland). (n.d.). *The educational philosophy*. Ermatingen: Author.

Urabe, K., Child, J., & Kagono, T. (Eds.). (1988). *Innovation and management: International comparisons*. New York: de Gruyter.

Useem, M., & Karabel, J. (1986). Pathways to top corporate management. *American Sociological Review, 51*, 4-200.

Van de Ven, A. H., & Ferry, D. L. (1980). *Measuring and assessing organizations.* New York: John Wiley.

Van Houten, D. R. (1987). The political economy and technical control of work humanization in Sweden during the 1970s and 1980s. *Work and Occupations, 14*, 483-513.

Varadarajan, P., & Futrell, C. (1984). Factors affecting perceptions of smallest meaningful pay increases. *Industrial Relations, 23*, 278-286.

Vollmer, F. (1986). Why do men have higher expectancy than women? *Sex Roles, 14*, 351-362.

von Clausewitz, C. (1968). *On war.* Middlesex, England: Pelican (Penguin). (Original work published 1832, translation published by Routledge & Kegan Paul, 1908)

Wagner, D., Ford, R. S., & Ford, T. W. (1986). Can gender inequalities be reduced? *American Sociological Review, 51*, 47-61.

Waldman, D. A., & Avolio, B. J. (1986). A meta-analysis of age differences in job performance. *Journal of Applied Psychology, 71*, 33-38.

Walker, G., & Weber, D. (1984). A transaction cost approach to make-or-buy decisions. *Administrative Science Quarterly, 29*, 373-391.

Walsh, J. P. (1989). Technological change and the division of labor: The case of retail meatcutters. *Work and Occupations, 16*(2), 165-183.

Warner, M. (1987). *Microelectronics and manpower in China* (Management Studies Research Paper, No. 5/87). Cambridge: Cambridge University, Engineering Department.

Werther, W. B., Davis, K., Schwind, H. F., Das, H., & Miner, C. F. (1985). *Canadian personnel management and human resources* (2nd ed.). Toronto: McGraw-Hill Ryerson.

Wheeler, L. (1986). Concluding comment. *Personality and Social Psychology Bulletin, 12*, 297-299.

Wholey, D. R. (1985). Determinants of firm internal labor markets in large law firms. *Administrative Science Quarterly, 30*, 318-335.

Williamson, O. E. (1985). *The economic institutions of capitalism.* New York: Free Press.

Willman, P. (1986). *Technological change, collective bargaining, and industrial efficiency.* Oxford: Clarendon.

Wilpert, B. (1986, July). Leadership and decision making in introducing new technologies. In J. Misumi (Chair), *Cross-cultural and interdisciplinary perspectives of leadership and organizational development.* Symposium conducted at the 21st International Congress of Applied Psychology, Jerusalem.

Windolf, P. (1986). Recruitment, selection, and internal labour markets in Britain and Germany. *Organization Studies, 7*, 234-254.

Young, R. L., Hougland, J. G., & Shepard, J. M. (1981). Innovation in open systems: A comparative study of banks. *Sociology and Social Research, 65*, 177-193.

Zicklin, G. (1987). Numerical control machining and the issue of deskilling. *Work and Occupations, 14*, 452-466.

Znaniek Lopata, H., Fordham Norr, D., Barnewolt, D., & Miller, C. A. (1985). Job complexity as perceived by workers and experts. *Work and Occupations, 12*, 395-415.

Zoltan, E., & Chapanis, A. (1982). What do professional persons think about computers. *Behaviour and Information Technology, 1*, 55-68.

Zuboff, S. (1982). New words of computer-mediated work. *Harvard Business Review, 60*, 142-152.

Author Index

Abernathy W, 19
Abrahamson M, 159
Ackerman P L, 231, 232, 233
Adams J A, 231, 232
Adams J S, 209, 210
Adams S, 35, 38
Ambrose M L, 210
Anderson C S, 258
Andrews K R, 60
Ansen D, 200
Arnold H J, 87, 155
Arnold J D, 107
Arnould R J, 43, 59, 197
Ash R A, 105
Astley W G, 41
Attewell P, 229, 237, 239, 241, 242, 246, 266
Avolio B J, 124, 200
Awad E M, 78, 79
Azumi K, 177

Ballerstedt E, 219
Bandura S, 139
Barclay L A, 105
Bar-Hillel M, 106
Barnewolt D, 209
Baron J N, 202, 207
Barrett G V, 157
Bass A R, 105
Bateman T S, 161
Beatty R W, 111, 122, 123
Beck E M, 155, 206, 207
Becker G S, 146, 206
Behling V W, 124
Bell N E, 214, 218
Belous R S, 259
Belt J A, 105
Ben-Abba E, 106
Bengelmann D, 219

Ben-Shakhar G, 106
Berg I, 231, 232
Berger D E, 181, 199, 201
Bernardin H J, 111, 122, 123
Betriebsrat, 120
Bettis R A, 197
Bhagat R S, 179, 181, 183, 198, 262
Biddle B J, 117
Dielby W T, 202
Bielenski H, 219
BIGA, 241, 286, 291
Bigoness W J, 211
Bikson T K, 130, 139, 199, 202, 216, 264, 282
Bilu Y, 106
Birkwald R, 60
Bjorn-Andersen N, 238, 245
Blalock H M, 275
Blau P M, 69, 274
Blyton P, 97
Boulet J, 95
Boulian P V, 33
Boylan M, 195
Braithwaite R B, 274
Brand J F, 210
Bräuninger U, 216
Braverman H, 229, 235, 236, 238, 246
Breakwell G M, 181, 199, 202
Breaugh J A, 104
Bridges W P, 62
Briggs S R, 110
Brousseau K R, 273
Brumlop E, 187, 224
Büchteman C F, 219
Buitendam A, 26
Bureau of Labor Statistics, 95
Burke M J, 139, 145
Burns T, 71
Burton R M, 62, 269

Caldwell D F, 157
Cameron K S, 60
Campbell A, 246
Campbell J P, 121, 138
Card S K, 217
Carpenter B N, 110
Carroll S J, 121, 122
Carter N M, 243
Carter S B, 138, 272
Cascio C F, 78, 79
Cascio W F, 197
Casey B, 185, 238
Center for Public Resources, 141, 143
Chan L M W, 284
Chapanis A, 202
Chen J A, 124
Child J, 178, 256
Chusmir L H, 201
Clarke F H, 85
Clausen J A, 214, 218
Cohen Y, 76, 91, 146, 149
Cooper C L, 207
Cooper E A, 136, 157, 210
Cooper McGovern L, 264
Crosby F J, 211, 212

Darrough M N, 85
Das H, 78, 89
Davidson C, 206
Davies A, 191
Davis K, 78, 89
Davis-Blake A, 157, 159, 163, 273
Day R D, 139, 145
DDI, 142
Deci E L, 86, 170
de Pay D, 254
Desai A, 143, 146
Dewar R D, 62
Dickel K E, 43, 78
Dierkes M, 33, 213, 283
Dobbins G H, 124
Dodd N G, 183
Doswell A, 39, 191, 209, 213, 256, 257, 267
Drazin R, 31, 80, 83
Driver M J, 134, 136, 197
Droege C, 34, 173, 262

Dubin R, 80
Dubnoff S, 168, 237, 245
Duchin F, 139, 140, 143, 187
Dunnette M D, 121, 138
Dutton J E, 62
Dyer L, 154

Eason K, 238, 245
Edwards L N, 155, 207
Ellis P, 187, 191, 192, 265, 271
Emery F E, 262, 267, 285
Emmerij L, 241
Eriksson I V, 243
Ettlie J E, 62, 256

Farren C, 135
Fay J A, 94
Fedor D, 210
Feldman D C, 116
Feldman J M, 124
Fellman T, 216
Ferris G, 210
Ferry D L, 71
Festinger L, 210
Feuer M, 143, 146
Feuer M J, 146
Fife-Schaw C, 181, 199, 202
Fiorito J, 91
Fishbein M, 160
Fitch M K, 155
Fligstein N, 194
Floyd S W, 282, 293
Flug A, 106
Flynn D M, 62
Ford R S, 206
Ford T W, 206
Fordham Norr D, 209
Form W, 198, 213
Fossum J A, 155
Foster L W, 62
Frenzel U, 219
Friedman A, 185
Friedman N S, 137, 150
Friesen P H, 22, 24, 43, 47, 51, 54, 55
Futrell C, 87

Author Index

Gagliardi P, 85
Galbraith C S, 62, 65
Gannett B A, 124
Gattiker U E, 31, 32, 33, 34, 46, 60, 77, 79, 135, 138, 145, 158, 161, 162, 181, 183, 184, 199, 201, 208, 209, 210, 211, 212, 213, 214, 217, 218, 220, 241, 256, 264, 267, 273, 282, 284
Gibbs B, 77, 220, 281
Gielen U P, 181
Gierer R, 216
Gifford W E, 124
Glick H, 141, 146
Glick H A, 146
Goethals G G, 210
Gold B, 195, 197, 216
Goldstone J A, 43, 44, 176, 265
Goodman P S, 210
Gordon G C, 202, 216, 228, 243
Goryunov I Y, 18, 51
Grandjean E, 216
Greenberg J, 209
Greenwood R, 61
Griffin R W, 160, 161
Guion R M, 107
Gutek B A, 60, 130, 139, 181, 198, 199, 201, 202, 209, 213, 216, 217, 218, 230, 264, 282

Haavio-Mannila E, 155, 209
Hackman J R, 135, 239
Hage J, 274, 276
Halaby C N, 206
Hall R E, 119, 272, 274
Hall R H, 94, 231, 232, 235, 236, 237, 239, 240
Hall R J, 202
Hambrick D C, 43
Hammer T H, 135
Hansson R O, 110
Harris and Associates, 264
Head T C, 161
Hedge A, 264, 265
Hefti R, 136
Heneman H G III, 153, 154
Herr B M, 124

Herzenberg S, 158, 168, 186, 245, 291
Hickson D J, 34, 177, 258, 261
Hinings B, 61
Hinings C R, 34, 177
Hinrichs J R, 138, 264
Hirschheim R A, 260, 280
Hitt M A, 18, 51
Hlavka A, 199, 264, 273
Hodson R, 207
Hofstede G, 180, 181, 182, 194, 262
Hogan R, 110
Hogarth R M, 35, 197
Hollenbeck J R, 211
Homans G C, 274
Horvath D, 34, 177
Hoskisson R E, 62, 65
Hougland J G, 59, 61, 282
Howard A, 107
Hrebiniak L G, 30, 47, 50, 60, 261, 282
Hulin C L, 40

Iacono S, 238
Ichniowski C, 129
IG-Metall, 133, 190
Ilgen D R, 124, 211
Index Group Inc, 268
Ireland R D, 18, 51
Isabella L A, 119

Jacklin C N, 267
Jackofsky E F, 96
Jacoby S M, 158
Jain H C, 196
James L R, 33
Jones J G, 292
Jones J W, 138
Joyce W F, 30, 261, 282
Juergens U, 187, 224

Kagono T, 256
Kahn R L, 32, 35, 38, 238
Kalleberg A L, 231, 232
Kamm J B, 53, 280
Kandolin J, 155, 209

Karabel J, 194, 199
Katz D, 35, 38, 238
Kaufman H G, 200
Kauppinen-Toropainen K, 155, 209
Kaye B L, 135
Keen K, 77, 220, 281.
Kemp A A, 155, 206, 207
Kerr J L, 162
Kerr S, 86, 152
Kimberly J R, 17, 18, 47, 111
Kingstrom P O, 124
Kirmeyer S L, 162
Klein K J, 202
Kleingartner A, 258
Klimoski R J, 106
Kling R, 238
Klitzman S, 202, 216, 228, 243
Knipper H-J, 99, 290, 292
Koberg C S, 201
Kohn M L, 231, 238, 242, 245, 254, 273, 288
Kotter J P, 134
Kraft P, 168, 237, 245
Krahn H, 267
Kram K E, 118, 119
Krebsbach-Gnath C, 219
Kuechler M, 183
Kugler P, 155, 208
Kuhn S, 158, 168, 186, 245, 291
Kuhn T S, 258
Kulik C T, 210

LaLiberte M, 202
Landau J, 135
Landau K, 105
Landy F J, 210
Langseth P, 217
Larwood L, 60, 135, 145, 161, 181, 198, 199, 208, 209, 213, 214, 217, 218, 256
Lavallee L, 95
Lavelli M A, 138
Lawler E E, 83, 91, 121, 138, 160
Lawler J, 211
Lazarus R S, 33
Leblebici H, 258
Lee T, 181, 199, 202
Leibowitz Z B, 135

Lengnick-Hall C A, 43
Leontief W, 139, 140, 143, 187
Levin M H, 140, 288
Levine E L, 105
Lieberman M A, 260
Liker J K, 217, 218, 223, 246
Link N A, 18, 83, 258
Lippitt G L, 217
Locke E A, 33, 121
Long R J, 216, 255, 260
Lorence J, 208
Loscocco K A, 190
Louis M R, 118
Lowe G S, 267
Lowman C, 91
Lucas R, 77, 220, 281
Luthans F, 183

Mainstone L E, 124
Mankin D, 216, 263
Mankin D A, 130, 202, 282
Mansfield R, 69
March J G, 31, 38
Mason N A, 105
Mason R O, 43, 78
Masters M, 210
Maurice M, 34, 178, 184, 185, 195, 268
McCaul H S, 183
McClelland D C, 200
McKenna J F, 210
McMillan C J, 34, 177, 261
McMillen D B, 198, 213
McQuaid S J, 179, 181, 183, 198, 262
Medoff J L, 94
Meyer J W, 178
Meyer M W, 58, 68, 262
Miles R, 41
Miller C A, 209
Miller D, 22, 24, 34, 43, 47, 51, 54, 55, 173, 262
Miller G A, 177, 178, 262
Miller H D, 95
Miller J, 231, 238, 242, 254, 273, 288
Miller K A, 231, 238, 242, 254, 273, 288
Mincer J, 206
Miner C F, 78, 89
Minor F J, 136

Author Index

Mitchell D J B, 158
Mitroff I I, 29, 60, 79
Montgomery M, 96, 156
Moran T P, 217
Morgall J, 202, 213, 240, 268
Morgan G, 61
Möschel W, 254
Mossop J, 217
Mount M K, 124
Mountain D C, 284
Mowday R T, 33
Mueller C W, 137, 207
Murphy K R, 124

Neal A C, 185, 192, 196
Nelligan T, 34, 199, 200, 201, 213, 273, 282
Nelson F D, 91
Newell A, 217
Newton R D, 216
Nolan, Norton & Co, 268
Nord W R, 20, 24, 26
Nottenburg G, 210
Nuti D M, 156, 157, 159
Nutt P C, 216
NZZ, 95, 96, 111, 112

Obel B, 62, 269
OECD, 95, 140
O'Hara R P, 135
O'Keefe R D, 62
Okubayashi K, 246, 264, 269
Oldham G R, 135, 210, 239, 264
Olian J D, 93
O'Reilly C A, 157
O'Reilly A, 41, 59, 197
Oskamp S, 213
Osterloh M, 192
Osterman P, 76, 77, 236, 237
Ostroff C, 211

Panko R R, 191, 230, 256
Parcel T L, 207
Pava C, 40
Payne R L, 292
Pearce J L, 154, 157

Pennings J M, 26, 28, 30, 31, 261
Penrose J, 201
Perry J L, 154, 157
Perry L T, 52, 177, 221
Petermann T, 216
Peters L H, 96
Pettigrew A, 34, 86
Pfeffer J, 17, 36, 37, 38, 42, 46, 70, 76, 77, 120, 146, 149, 154, 157, 159, 160, 163, 211, 258, 262, 273, 295
Pfister D, 259
Piore M J, 99, 245
Porter L W, 33
Porter M E, 50

Quan N T, 87, 159

Rafaeli A, 106
Ranson S, 61
Rauschenberger J M, 107
Ray C A, 233, 234
Reich M, 206
Reilly R R, 106
Reith J, 138
Robertson I T, 207
Robey D, 238, 245
Rodriques S, 161, 244
Rogers T F, 137, 150
Rohmert W, 105
Roitman D B, 217, 218, 223, 246
Rokeach M, 273
Rollier M, 186, 189
Rosegger G, 18, 20, 22, 195
Rosenbaum J E, 119, 126, 134
Roskies E, 217, 218, 223, 246
Ross J, 120, 273
Rotchford N L, 264
Rowe A J, 43, 78
Roznowski M, 40
Rubenson K, 140, 241, 292
Ruch L, 219
Rynes S L, 93

Sabel C F, 245
Salancik G R, 36, 42, 46, 120, 154, 258, 262

Salter J R, 96
Sandholtz K W, 52, 177, 221
Scarbrough H, 192
Schein E H, 83, 135, 230
Schneier E C, 121, 122
Schoenbach C, 231, 238, 242, 254, 273, 288
Schoenberg R, 231, 238, 242, 254, 273, 288
Schoenherr R A, 69
Scholl R W, 210
Schooler C, 231, 238, 242, 254, 273, 288
Schoonhoven C B, 28, 31
Schreyögg G, 29, 31
Schuetze H G, 140, 288
Schwab D P, 153, 154
Schwartz J E, 274
Schweizerischer Gewerkschaftsbund, 241
Schwind H F, 78, 89
Schwitter J P, 34, 177
Scott W R, 36, 69, 178, 262
Seiford L M, 201
Sekaran U, 183
Selig G J, 260
Selznick P, 60
Sethia N K, 85, 86
Shaiken H, 158, 168, 186, 245, 291
Shapira Z, 160
Shepard J M, 59, 61, 282
Shepelak N J, 211
Sheridan J E, 116
Sherman J D, 87
Shirom A, 162
Simon H A, 31, 35, 38, 60, 283
Smith H L, 87
Snow B R, 202, 216, 228, 243
Snow C C, 41, 47, 50, 60
Sorge A, 34, 178, 184, 185, 186, 187, 195, 236, 245, 246, 255, 268
Soubel W G, 107
Spencer J, 181, 199, 202
Spenner K I, 229, 231, 232, 234, 254, 266
Spulber D F, 197
Stalker G M, 71
Stark D, 76, 77, 162
Staw B M, 120, 212, 273
Steers R M, 33
Stellman J M, 202, 216, 228, 243
Stepina LP, 210
Stevenson W, 58, 68, 262

Stevenson W B, 154, 157
Stevenson H H, 60
Stinchcombe A L, 159
Stromberg A H, 181, 198, 199
Swaroff P G, 105
Swidler A, 34, 175, 193, 194, 220, 240, 262
Sydow J, 34, 216, 257

Tachibanaki T, 194, 207
Tanaka H, 193, 194, 200, 201
Tassey G, 18, 83, 258
Taylor R P, 138
TCO, 272
Tenopyr M L, 106
Terborg J T, 95
Tetrick L E, 33
Thompson D E, 124
Thorndike E L, 139
Tiedeman D V, 135
Tolliver J M, 124
Tornatzky L G, 138
Treacy M E, 268
Triandis H C, 179
Trist E L, 262, 267, 285
Troy N, 219
Tubbs M E, 121
Tucker S, 20, 24, 26
Turner J A, 244

UBS, 97, 156, 159
Urabe K, 256
Useem M, 194, 199
Utterback J, 19

Vancouver J B, 211
Van de Ven A H, 31, 41, 71, 80, 83
Van Houten D R, 189, 190, 243
Varadarajan P, 87
Vollmer F, 267
von Clausewitz C, 78
Von Glinow M A, 85, 86
von Thienen V, 33, 213, 216, 283

Wagner D, 205

Waldman D A, 124, 200
Walker G, 197
Walsh J J, 260
Walsh J P, 242, 243, 246
Warner M, 10, 34, 178, 184, 185, 186, 187, 195, 246, 255, 268
Wayne S J, 161
Weber D, 197
Webster S, 58, 68, 262
Weick K E Jr, 121, 138
Werther W B, 78, 89
Wheeler L, 210
Whetten D A, 60
Wholey D R, 76
Williamson O E, 46, 61, 62, 94, 254
Willman P, 191, 203
Wilpert B, 34, 203, 268
Windolf P, 94, 133, 155, 162, 252, 272
Wright Kassner M, 210

Young R L, 59, 61, 282

Zedeck S, 106
Zicklin G, 241, 254
Znaniecki Lopata H, 209
Zoltan E, 202
Zuboff S, 269
Zuercher, 137

Subject Index

Absolute surplus value, 237
Alternative employment strategies, 95-99
  flexible work scheduling, 99, 103
  guaranteed part-time employment, 95-96
  nonguaranteed part-time employment, 96-97
  overtime, 99, 102
  part-time employment, 95-97, 100
  term positions, 98-99, 100-2
  work sharing, 97-98, 102
Analyzer strategy, 54
Apple Macintosh, 201, 243
Appraisal, 114-132
  Behaviorally anchored rating scales (BARS), 123
  categorization, 124
  critical incident, 122
  forced distribution, 122
  paired-comparisons, 122
  performance appraisal process, 121
  rank ordering, 122
  Source and Accuracy of Data, 123-125
  Trait approach, 122
  use of data, 125
Automating production, 18

Behaviourally anchored rating scales (BARS), 123
Betamax, 51
Brazilian. *See* Brazil
Bureaucracy, 58

Canadian. *See* Canada
Career
  career development system, 133-151
  career management, 78
  career satisfaction, 134
  centralized data bases, 136
  lateral mobility, 135
  mobility, 134, 146
  resource dependence model, 135, 157
  Strategic contingency theory, 135
Chinese. *See* China
Common Market, 268, 291
Comparable worth, 157-158
Compensation systems, 86-88, 152-171
  remuneration, 83, 87-88, 152
  rewards, 84, 87-88, 152
Computer-aided manufacturing, 18
Computer-numeric control, 186, 241
Contingency, 28-39
Contingency theory, 28, 29, 31
Cross-cultural, 28, 34
Cross-national differences, 245
Culture, 28-39, 175-204, 205-9, 240-247
"Culture-free" hypothesis, 177

Declarative knowledge, 231
Defender strategy, 51-52
Deskilling, 232-235, 241
Differentiation, 51, 54
Diffusion, 21-4, 26-27, 41, 54-55
Discrimination, 155, 205-227 ✓
Distinctive competencies, 60-61
Dutch. *See* Netherlands

Educational Leave Act, 140
Emerging or changing technology culture, 244
Emery, 53
Employment strategies, 93-113
End-user, 23, 37, 39, 228
Environment, 16, 33
Environmental determinism, 15, 16, 40-56, 74, 82

Subject Index

Ergonomics, 215-216
Established market, 242
European Community, 268, 291
EUTM, 228-240

Fleet Street, 119, 137
Fortune 500, 194, 201
Free Trade Agreement, 182, 291

German. *See* Germany
Governance, 60
Greens, 25

Human resource, 93, 111
Human resource culture, 83-86
Human resource management (HRM). *See* SHRM

IBM-PC, 201
In-house training, 142-145, 158
Innovation
   derivative innovation, 21-22
   incremental innovation, 24, 25, 54
   primary innovation, 21-22
   process innovation, 19-20
   product innovation, 19-20
   radical innovation, 24-25, 54, 40
   technological innovation, 25-27
Innovation diffusion, 22-24
   change in the innovation, 23
   complementaries among innovations, 23
   origin of the innovation, 22
Internal labour market (ILM), 94, 104, 153, 158
   allocative dimension, 75
   determinism (ILMD), 29, 30, 34, 74-94, 114-115, 133, 157
   economic sector, 76
   organization size, 76
   unionization, 76
Internal labour market culture, 89
Internal structures and processes
   authority structures, 69
   communication systems, 70-71

   decision-making and negotiating strategies, 69-70
   distinctive competence, 60-61
   governance, 60
   nature of tasks, 66-68
   organization of work, 68-69
   ownership, 59
Introduction of innovation
   discontinuous, 44, 45, 52, 53
   evolutionary, 23, 44, 45, 52, 53

Japanese *See* Japan
Job classification system, 90
Job content compensation
   job content, 160-162
   social information processing model, 160
Labour goals, 89 90
Labour relations, 78
Learning, 138-139
Librium, 65
Life-long learning. *See* Recurrent Training
Local-Area-Network (LAN), 144
Lotus, 85

Macro, 28, 35, 37
Malayans. *See* Malaysia
Mature market, 242
Mature technology culture, 245
Mechanization, 18
Mentor, 118-119
Meta-analysis, 121
Micro, 28, 35, 37
Multiplicative interaction, 31

Natural selection, 43, 46

Open systems, 58
Organizational choice, 15, 40-45
Organizational structure
   lateral mobility, 80
   multidivisional (M-form), 61-62, 68, 70
   personality assessment system, 80
   promotion ladders, 80, 82
   reward and remuneration system, 80

Organizational structure (*continued*)
  selection system, 80
  skill specificity, 80
  specialization, 80
  unitary (U-form), 61-2, 68, 70
Organizational adaptation, 40-56
Organizational culture, 83, 155, 220-224
Organizational effectiveness, 69

Part-time employment, 95-97, 100, 158
Pay compensation, 152-159
  ILMD and pay levels, 157
  merit pay systems, 154
  pay, 153-154
  pay levels, 153
  pay structures, 154, 158-9
  pay system, 154
  profit-sharing programs, 156
  two-tier wage structures, 158
Perceptions, 31, 33, 34
Perestroika, 176
Performance appraisals, 114, 121-123
Proactive, 40
Proactive adaptation, 41, 42, 44, 53, 57
Proactive change, 41
Procedural knowledge, 231
Processes, 80, 81
Prospectors, 47

Rating scale, 122
Reactive, 41, 53
Reactive adaptation, 41, 42, 53, 57
Reactive change, 41
Reactor strategy, 51
Recruiting
  headhunter, 109
Recruitment, 93-113
  ideal candidate, 104
  status quo strategy, 107
Recruitment and selection systems, 93-113
Recurrent training, 139, 241
Relative surplus value, 237
Reliability, 106, 123
Remuneration. *See* Compensation
Resource dependence model, 157, 135

Reward system, 87-88, 152
Risk management, 197-198

Selection
  graphology, 106
  job analysis, 105
  personality testing, 106
Selection systems, 93-113
Skilled workers, 236
Skills, 228-255
Socialization, 114-132, 138, 149
  anticipatory socialization, 116
  change and acquisition phase of socialization, 116
  encounter phase of socialization, 116
  implementing, 118-20
  interpreting events on the job, 117-118
  mentors, 118, 119
  work roles, 117
Socialization systems, 114-132
Spanish. *See* Spain
Specialization, 18
Stage strategies for adapting technology
  cost leadership, 51
  defender, 51, 52
  differentiation, 51, 54
  focuser, 51
  reactor, 51
Stakeholders, 29, 36, 57, 63
Standardization, 18
Status differentiation compensation
  status differentials, 163
  status differentiation, 152, 162-164
Strategic choice, 40-56
Strategic human resource management (SHRM), 29-31, 34, 75-92, 93-94, 114-115, 197
Strategic management, 42-43
Structural contingency theory, 80
Subjective culture, 193-198
Swedish. *See* Sweden
Swiss. *See* Switzerland

Talent additive, 159
Talent complementary, 159

Subject Index 335

Term positions, 98-99, 100-102
Tien An Men Square, 176
Training, 133-151
   association, 139
   extra-organizational training, 141-142
   imitation, 139
   in-house training, 142-145, 158
   nonmanagerial employee training, 144
   recurrent comprehensive education and training, 139
   seniority university, 141
   supervisory and managerial training, 145
   Young employee training, 143
Tylenol, 29, 36

Union, 74-76, 89
Unionization, 91

Validity, 106, 123
   predictive, 106
Valium, 65
VCR, 51
VHS, 51

West German. *See* Germany
Work flow, 188-92
Work sharing, 97-98, 102

Places Mentioned in Text

America. *See* U.S.

Baltic Republics. *See* Estonia; Lithuania
Brazil, 243

Canada, 25, 30, 47, 54, 63, 66, 78, 95, 98-99, 140, 144, 156, 181, 182, 196, 199, 208, 210, 286, 288, 290, 291
Chernobyl, 228
China (PRC), 44, 176, 195, 245
Czechoslovakia, 189

Denmark, 244, 290

England. *See* Great Britain
Estonia, 177
Europe, 23, 25, 26, 45, 47, 95, 102, 106, 140, 141, 143, 176, 198, 258, 264, 267, 290

Finland, 155, 207
France, 65, 178, 184, 194

German Democratic Republic (GDR), 176
Germany, West. *See* West Germany
Great Britain, 44, 50, 94, 119, 176, 184, 191, 195, 199, 202, 245
Greece, 290

Holland. *See* Netherlands

Italy, 156, 189, 190

Japan, 115, 144, 155-156, 183, 185, 189, 192, 194, 196, 200, 206, 245, 264

Korea, 189

Latvia, 177
Lithuania, 177

Malaysia, 181
Mexico, 290

Netherlands, 96-97, 109, 287, 288
New Zealand, 63
North America, 23, 47, 52, 146, 155, 188, 200, 237, 257, 264, 271, 285
Norway, 185

Paris, 65
People's Republic of China (PRC). *See* China

Soviet Union. *See* USSR
Spain, 99, 289, 290, 291
Sweden, 140, 153, 188, 189, 190, 192, 271, 287, 288
Switzerland, 20, 22, 50, 65, 66, 95, 137, 143, 144, 155, 182, 200, 207, 218, 240, 286

Turkey, 44

Union of Soviet Socialist Republics (U.S.S.R.), 177, 180, 195, 228
United Kingdom. *See* Great Britain
United States (U.S.), 25, 26, 29-30, 50, 52, 54, 65, 94, 95, 96, 115, 136, 137, 140, 143, 154, 156, 181, 183, 185, 192, 194, 198, 199, 201, 206, 210, 245, 290

West Germany, 26, 50, 73, 94, 99, 144, 155-156, 176, 181, 184, 187, 192, 195, 218, 237, 270, 287, 289

Subject Index

Companies and Associations Mentioned in Text

Academic Press, 268
Airbus 320, 42, 170
Air Canada, 30, 288-9
Air France, 42, 170
Air Inter, 45, 170
Alberta Government Telephones (AGT), 66
Amadeus, 45, 52
American Airlines, 26, 54, 270
American Automobile Workers Union, 115
Apple Macintosh, 201, 243
Arthur Andersen & Co., 128, 142, 145
Ashton-Tate, 85
AT&T, 143

Bank of America, 164, 171
Bank of Montreal, 86
Bank of Tokyo, 145
British Airways, 45
British Leyland (BL), 192
British Telecom, 50
Burlington Northern, 82, 153

Canada Post, 54, 69, 85, 87, 102, 103, 109, 126, 129, 148
Canadian Airlines International, 30, 164
Chrysler, 179
Ciba-Geigy, 86
Cirrus, 53
Citibank, 187, 268
Continental, 270
Credit Suisse, 20, 54, 55, 89, 102, 109, 136, 148, 252

Dawoe, 145
Deutsche Bundespost (DP), 26, 50, 71
Development Dimensions International (DD), 142, 145
Dow Chemical, 149

Eastern Airlines, 54, 271
Exxon Valdez, 229

Federal Express, 26, 53
Fiat, 186, 187, 198-190, 242-243, 262
First Interstate, 53
Fletcher Challenge, 63
Ford, 94
Frontier, 270

Galileo, 45, 52
General Motors (GM), 62, 63, 65, 94
German Bundespost (German PTT). *See* Deutsche Bundespost
German Metal Manufacturing Association (IG-Metall), 99, 103, 111, 133, 190

Harcourt Brace Jovanovich, 187
Hoffman La Roche, 65
Hollandsche Beton Groep, 69

Iberia, 45
IBM, 18, 23, 136, 145, 149, 201
ICI, 86
IG-Metall. *See* German Metal Manufacturing Association

Johnson & Johnson, 29, 36

KLM, 45

Louis Vuitton-Moet Hennessy, 65, 71
Lufthansa, 45

Marathon Oil, 52
Matsushita, 82, 144
Mövenpick, 66-68, 69, 167-169

NASA, 23
Nestlé, 145

New York Air, 270
Nissan Corporation, 115
Nixdorf Computer AG, 259

People Express, 270
Peugeot, 179
Procter & Gamble, 88, 129
Purolator Courier, 288

Saab-Scania, 188-190
SABRE, 26, 45, 52, 54, 270
SAS, 45
Schwab Securities, 163-164, 171
SEAT, 99, 290
Siemens, 86, 128, 129, 148, 186
Sony, 51, 82, 109
Sterling Drug Inc., 65
Swedish Metal Workers' Union (SIF), 153, 162
Swissair, 42, 45, 170
Swiss Post, Telephone and Telegraph (PTT)
Swiss PTT, 50, 71, 88, 103, 109, 130, 148, 253

Tages Anzeiger, 137
Tecan, 259
Texas Air Corporation, 54, 270
Toyota, 25

Union Bank of Switzerland, 53, 145
United Parcel Service, 288
United Airlines, 45, 54
U.S. Sprint, 50
U.S. Steel, 52

Verband der Automobilindustrie (VDA), 73
Volkswagen, 99, 186, 192, 290
Volvo, 188-190, 239, 242, 262

Wardair, 30
Wild-Leitz, 259
WordPerfect Corp., 197, 269
World Airways, 100

About the Author

URS E. GATTIKER is Associate Professor of Organizational Behaviour and Technology Management in the Faculty of Management at the University of Lethbridge, Alberta, Canada. He received his Ph.D. in Management and Organization from Claremont Graduate School, in the United States, in 1985. Before entering academia, he worked in Switzerland, South Africa, and the United States. He is spending the 1990-1991 academic year on sabbatical at Sanford Graduate School of Business. He is editor of *Technological Innovation and Human Resources*, a biannual book series, and founding editor of *Technological Studies* (*TS*), an interdisciplinary journal with its first issue scheduled to appear in 1992. His research interests currently include skill acquisition and human capital theory, technological change, career development, and quality of work life issues. He is Chair of the Academy of Management's Research Methods Division. During the writing of this book, he was a member of the executive of the Academy of Management's Technology and Innovation Management Division.